MAN

HIS FIRST TWO
MILLION YEARS

MAN

HIS FIRST TWO
MILLION YEARS

A BRIEF INTRODUCTION
TO ANTHROPOLOGY

Ashley Montagu

A DELTA BOOK

To Barbara

A DELTA BOOK

Published by Dell Publishing Co., Inc.
750 Third Avenue, New York, N.Y. 10017
First published under the title
Man: His First Million Years
Delta ® TM 755118, Dell Publishing Co., Inc.
Published in hardcover by Columbia University Press
Printed in the United States of America
Second Printing

PREFACE

THIS BOOK HAS BEEN WRITTEN for people who want to learn, no matter what their chronological age may be—people who find their interest in life steadily increasing and who expect to find it continuously so. By "interest" I do not mean the enthusiasm of the moment but the intense and enduring interest of a lifetime.

It seems to me that anthropology should form a part of the intellectual equipment of every person who enjoys the benefit of a high school education. Indeed, I believe that anthropology should form the core of the educational curriculum at all levels, grammar school, high school, and college. There are now a number of high schools in which anthropology is taught as a regular course, and there are also some private schools in which it is taught—with great success. It is to be hoped the number of these schools will greatly increase. Meanwhile, those who desire a simple, uncomplicated introduction to anthropology may find the present book helpful.

Human beings are, surely, the most interesting creatures on this earth, yet, strangely enough, their study has been left to almost the very last of the sciences to come into being. Anthropology, the science of man, is just about the youngest of the sciences, and it is, without any doubt, the most important —the most important because it deals with humanity itself, with the great questions of life and of death. What is man? What is he born as? What is he born for? Where did he come from? Whither has he been going? How did he come to present so many different types, both physically and culturally? These, and many more, are some of the questions which anthropology asks and attempts to answer. There can be scarcely any questions more important than these, unless it be the question: In what direction should man go? Here, I think, anthropology is in a position to return some fundamental answers, for if one would know what man is born *for,* one ought to begin by learning what he is born *as.* It is equally important to understand something of the routes man has taken to achieve the remarkable versatility that characterizes him.

There are some anthropologists who declare that it is not the business of anthropology to tell anyone what he was born for. The scientist, they say, should not be concerned with value judgments, with distinguishing between what *is* and what is *desirable* and then going on to recommend the desirable. The scientist's job, they claim, should be to discover what *is;* as for what *should be,* that can be left to the individual or philosophers. It is further argued that one cannot draw any inferences or seek a direction from what is for what should be. This is a point of view with which I do not agree. I think one can and should study man in order to learn what he is so that we may the better be able to make him what he should be. Knowledge for the sake of knowledge is all very well, but where man is concerned we need to apply the findings of science to the problems he creates. That way we may eventually accomplish the realization of the best that is in everyone, for the maximization of their own and their fellow human beings' happiness.

I have developed this point of view, in the light of the scientific evidence, in my book *The Direction of Human Development* (Hawthorn Books, New York, 1970). In the present volume I have adhered principally to the statement of the facts of anthropology. The reader will, I expect, be able to go on from them to the proper conclusions. And so let us begin.

Ashley Montagu

2 November 1968
Princeton, N.J.

CONTENTS

MAN

HIS FIRST TWO
MILLION YEARS

❧ 1 ❧

INTRODUCING

OURSELVES

Mᴀɴ, *Homo sapiens,* meaning "man the wise," is the most interesting creature on the face of the earth. At least, he is so to himself—anthropocentrically, the most interesting, the most exciting, the most promising, and a great deal more. The term "man" is here employed generically to include both sexes, in all their developmental stages, from embryo to old age.

How has it come about that man should be the most interesting, the most exciting, and the most promising of all the creatures on this earth? Is it not rather conceited to speak of ourselves in such a manner? There are some who believe that it is. Some there are who think that man is a vulgar, oafish upstart who, with supreme arrogance and in the worst possible taste, has awarded himself the highest of all accolades, the name *Homo sapiens.* They feel that the appellation *sapiens,* the wisdom part, has yet to be earned, and that for civilized man, at any rate, the proper appellation at the present stage of his development would be *Homo sap.* or "the wise guy," one who has grown far too clever for his own good. It should, however, be added that such critics do not deny that man possesses the necessary potentialities for wisdom. What they do deny is that he has thus far made the best of those potentialities. There have been some so cynical as to suggest that modern man in fact represents the long sought "missing link," that he is, in fact, "the missing link" between the ape and true man. Cynicism, it has been remarked, is the last refuge of an idealist. It is, perhaps, somewhat premature to express disappointment that man is not yet what he is potentially capable of becoming, for compared with other living creatures he is very young indeed. He stands in need of more time and more understanding.

The name *Homo sapiens* ("Homo" stands for the zoological *genus,* and "sapiens" for the zoological *species*) was first given to man as a zoological group in 1735 by the great Swedish botanist and systematist Karl von Linné (1707–1778), known to the world, and later referred to in this book, by the Latinized form of his name, Linnaeus. When he named man *Homo sapiens,* he wanted to make it quite clear that the distinguishing mark of man was the extraordinary ability with which he was able to use his mind, an ability in which the human species greatly surpasses all other creatures. In this sense we, as a species, are well named. Without any conceit it is very desirable for us to understand ourselves for what we are, for the better we know ourselves, the better we shall be able to realize our potentialities. As human beings we are the most interesting creatures in the world because we are the most interested. The best way to be interesting is to be interested in some thing or things; then you cannot help but be interesting to other people. There is no limit to the interest of human beings, an interest which is often called curiosity. Man is the most curious of all creatures. He is always wanting to find out. He is the Nosy Parker of Creation. With his curiosity he has conquered land, sea, and air; he has split the atom, tunneled the earth, compressed distances, created radio and television, traveled into space, and landed complex measuring devices on the moon.

Through his outstanding representatives, man has given us all the beauty of poetry, the drama, great books, painting, the dance—all art, indeed—great thoughts, and greatest of all, great lives, as those of Moses the founder of the Mosaic Law, the Indian religious leader Buddha (563?–483? B.C.), the Chinese sage Confucius (551?–479? B.C.), Christ (6? B.C.–A.D. 29?), the fountainhead of Christianity, and in our own time such men and women as Abraham Lincoln (1809–1865), Mohandas K. Gandhi (1869–1948), the great Indian leader, Albert Schweitzer (1875–1965), Martin Luther King, Jr. (1929–1968), the heroic American Negro leader, Florence Nightingale (1820–1910), "the Lady with the Lamp," modernizer of nursing and hospital administration, and Susan Anthony (1820–1906), the great reformer and leader of the woman suffrage movement.

This book, in fact, would not be large enough to hold all the names of those human beings who have earned, by their achievements, the title of "great.". Every people has had its great men, and the names of most of these have been forgotten. Recorded history is no more than six thousand years old, whereas human beings have been making history ever since they have been on this earth, a period of about two million years. The long period of human experience before recorded, or *written,* history begins, is known as prehistory or the prehistoric period. Very few great names have come down to us from that period, and yet we can be certain that there were men and women quite as great living in the periods of prehistory as there have been in the historic period. The works they have bequeathed to us are unsigned.

If we seem to know more than our ancestors did, it is not because we have brains that are more highly developed, but because we have inherited all that they have bequeathed to us in the form of knowledge and wisdom. We can see farther ahead and over a wider territory not because we have better eyes, but because standing, as it were, on their shoulders we can see more—but if it were not for their shoulders we would be able to see no more than they were able to see. Oftentimes we are not able to see as much, for, casting our vision over so broad an expanse of territory, we tend to miss the things that are closest to our noses. That is why, every so often, it is a good thing to hang a question mark on the things we take most for granted. Since some of the things men take for granted are likely to be wrong and some right, it is highly desirable that every society encourage the inquiring mind. If we are ever to distinguish between the right and the wrong, we shall most efficiently be able to do so through self-examination and inquiry. Happily, man is the most examining and inquiring creature in creation, so we need not worry. We should begin to worry only when obstacles are created to prevent men from pursuing their legitimate inquiries in freedom and without restraint.

Being as interested as he is in the world around him, man, from the earliest times, has learned to make use of his environments in so rich a variety of ways that were one to spend the whole of one's life trying to learn about them, one could acquire knowledge of only a small segment of them. Man has greatly added to his own intrinsic interest by the way in which he has improved upon or added to the environments in which he has found himself. But he is also interesting because of what the environments have done by way of improving him, producing all the wonderful varieties in which we find man today both as a physical and as a social creature. But, then, this is what this book is mostly going to be about.

That man is the most exciting of all creatures is, in part, due to his peculiar faculty for adventuring into unexplored territories. Now that so much of the earth has been explored, he is beginning to turn his attention to the exploration of outer space. The future of man in a world served by the harnessed atom is most exciting to contemplate. Mechanical brains, mechanical factories, the outmoding of muscle power, the increase of leisure, longer life, better health, peace and good will on earth unto all men—the horizons are unlimited! It is good, and the highest of all privileges, to be a member of the species *Homo sapiens*.

The promise of *Homo sapiens* is such that we have but to apply what we already know about his nature to make the world a very happy place for virtually everyone. *Homo sapiens* is the most educable of all living creatures—which means that he can learn quite as many wrong things about himself as right ones, and hence confuse himself very much more efficiently than any other creature. Man at the present time, in many parts of the earth, is seriously suffering from such confusion. And we men of the Western world are suffering from this confusion in many ways, perhaps as much as people anywhere else in the world, if not more so.

In order to make clear in our minds the causes of this confusion—and it is only by doing so that we will be able to extricate ourselves from it—it is necessary to learn what we can of the ways in which, and the routes by which, we came to be as we now are. The body of organized knowledge which deals with this subject is called anthropology, the science of man, surely the most interesting science that one could possibly study.

As Kaj Birket-Smith, the Danish anthropologist, has written, anthropology is "mankind's great means of self-education, since it tries to explain in order to create understanding, and thereby provides the opportunities for aiding development and putting it on the right track."

❧ 2 ❧

INTRODUCING
ANTHROPOLOGY

ANTHROPOLOGY is the science of man. That is a rather broad definition. Let us try to draw a line around it so that we may really grasp what the science is about. The word "anthropology" is derived from two Greek words, the one *anthropos* meaning "man," and the other *logos* meaning "ordered knowledge." So anthropology is the ordered knowledge of man. The fact is that anything relating to man is grist to the anthropologist's mill. Anthropology is divided into two main parts: (1) cultural anthropology and (2) physical anthropology.

CULTURAL ANTHROPOLOGY

Cultural anthropology is concerned with the study of man's cultures. By "culture" the anthropologist understands what may be called the man-made or learned part of the environment: the pots and pans, the laws and institutions, the art, religion, philosophy, mythology, and the like. Whatever a particular group of people living together as a functioning population have learned to do as human beings, their way of life, in short, is to be regarded as culture. The cultural anthropologist studies different cultures and compares them with one another in order to learn how it is that people come to do what they do in so many different ways, and also to learn, wherever possible, the relationships of one culture to another. He tries to find those common elements in all cultures which can be summarized in terms of generalizations or laws which are true of all cultures.

Where there are differences, he tries to find the causes of these differences.

The cultural anthropologist is interested in all the forms that human social behavior assumes in organized societies. He is, however, interested in very much more than the mere description of these forms, for he desires to understand how they came into being, and so the cultural anthropologist must often be quite as good a psychologist as he is anything else. Fundamentally what he is interested in is the nature of human nature. Today there is quite a flourishing school of anthropologists known as the personality-in-culture school. These cultural anthropologists, such as Margaret Mead, John Honigmann, and Francis Hsu, are interested in tracing the relationship of the cultures in which human beings are socialized, that is, brought up, to the kind of personalities they develop.

Other cultural anthropologists are interested as specialists in studying such aspects of the cultures of different peoples as their legal institutions, social organizations, religion, mythology, language, and their material culture such as their art, pottery, basketry, implements, and the like. Some cultural anthropologists take whole tribes for their special study and spend from many months to years attempting to understand every aspect of their culture.

Traditionally the cultural anthropologist has studied the so-called primitive peoples of this earth. In reality there are no "primitive peoples." The peoples so called are technologically not so highly developed as civilized peoples, but their kinship systems, languages, and other aspects of their culture may be considerably more developed or at least as well developed as those of civilized peoples. For these reasons most anthropologists prefer to speak of such technologically less developed peoples as "nonliterate peoples"—peoples without writing. The major part of the anthropologist's attention still continues to be devoted to the cultures of such peoples. In more recent years, however, anthropologists have been turning their attention to the study of the technologically more advanced peoples of the earth. Now we have good anthropological studies not only of the Australian aborigines and the Congo Pygmies, but also of the Japanese, the Chinese, the Germans, the Americans, the English, the Norwegians, and many other peoples.

Formerly the study of modern societies was left to the sociologist (sociology = the study of society). The methods of the cultural anthropologist have greatly influenced those of the sociologist, but the difference between the two disciplines remains. The sociologist studies modern societies in great detail, while the anthropologist is principally interested in the study of nonliterate cultures and brings to the study of modern societies a method which is at once wider and deeper than that of the sociologist.

Today there are specialists who are known as *applied* anthropologists. These are essentially cultural anthropologists who bring their special methods to bear principally upon the practical problems of industry, government, the professions, and the like. They go into a plant and study the relationships between the workers and their employers, between the workers and their work, and they advise on the methods of improving these relationships. They perform similar services in government and in other agencies.

There are cultural anthropologists who work in hospitals in collaboration with psychiatrists. They study the relationships within the hospital between patient and doctor, administration and staff, and they conduct collaborative studies of whole districts in order to throw light upon the genesis of mental illness and its possible prevention.

The collaboration between anthropologists and psychiatrists in the study of the cultures of different societies has been very fruitful indeed, and holds much promise for the future. Other cultural anthropologists have made notable contributions to the study of the cultures of poverty in modern societies.

Another branch of cultural anthropology is *archeology* (often unkindly called the moldier part of anthropology or "the Science of Rubbish"). Archeology is the science which studies cultures that no longer exist, basing its findings on the study of cultural products and subsistence remains recovered by excavation and similar means.

Since the cultural remains, the artifacts, represent the physical manifestations of a culture, and their distribution at a site tends to reflect something of the behavior of its members as well as their qualities, it is possible to reconstruct something of the social system of a culture, the way of life of a people that has long ceased to be.

If anthropologists are "the glamour boys" of the social sciences, archeologists are "the glamour boys" of anthropology. All of which means that their specialty can be a very exciting one indeed, even though it generally entails a great deal of hard work, with far too much dust and sand in one's hair, one's boots, and one's desiccated sandwiches. What the archeologist is interested in doing is not merely to disinter an extinct culture, but to trace its relationships to other cultures. In this way archeologists have been able to solve many problems that would otherwise have remained puzzling and to link up cultures that in their present form hardly seem related.

Anthropological archeologists are to be distinguished from classical archeologists. Classical archeologists are interested in the extinct cultures of classical antiquity, such as those of Greece, Rome, Crete, Persia, and Palestine, while anthropological archeologists are interested in the cultures of prehistory.

PHYSICAL ANTHROPOLOGY

Just as the cultural anthropologist is interested in studying man as a cultural being, so the physical anthropologist is interested in the comparative study of man as a physical being. The physical anthropologist studies the origin and evolution of man's physical traits and the diversity of forms which those physical traits may take. He tries to discover the means by which the likenesses and the differences between groups of human beings have been produced. He tries to discover how man, in his various physical shapes, came to be the way he is now.

Since cultural factors have played a highly important role in the physical evolution of man, the physical anthropologist must be well trained in cultural anthropology. The physical anthropologist must never lose sight of the fact that throughout his evolutionary history man's principal mode of adapting himself to his environment has been through culture.

There are all sorts of specialists in physical anthropology. There are those who study man's origin and evolution; these are called paleoanthropologists (*paleo* = old). They study the skeletal remains of fossil man and everything related to them. Then there are those who study the comparative anatomy of the primates, the classificatory group to which man belongs together with the apes and monkeys. Here the primatologist, as he is called, is interested in discovering possible physical relationships among the large variety of types that constitute this group.

There are physical anthropologists who study growth and development. There are those who study physiology, those who study the blood groups and blood types, and others who study biochemistry. All these studies are combined with the study of genetics, the science of heredity or how living things come to exhibit the likenesses and the differences they do. There are those who study the so-called races of man, and there are a few who study all these subjects in the attempt to unify what they know into a consistent body of knowledge.

There are also applied physical anthropologists. These are largely concerned with the measurement of man, anatomically and physiologically, in order to determine, for example, standards for clothing, equipment in industry or in the armed forces, hat sizes for men, shoe sizes for women, and the like.

There are also some constitutional physical anthropologists. These attempt to study the possible relationships between body-form and disease, body-form and personality, and even body-form and ethnic group.

ANTHROPOLOGY AND HUMANITY

The anthropologist is first and foremost interested in human beings, no matter what the shape of their heads, the color of their skin, or the form of their noses. As a scientist he is interested in the facts about human beings, and as an anthropologist he knows that the facts are vastly more interesting than the fancies, the false beliefs, and the downright distortions of the facts which some misguided persons seem always to have found it necessary to perpetrate. Humanity has a wonderfully interesting history, and it is the principal function of the anthropologist to reveal that history to the student as simply and as clearly as possible. This is what we shall now attempt to do in the following pages.

In concluding these introductory paragraphs I should like to quote the words of a great biologist, Thomas Henry Huxley

(1825–1895), writing in 1889 to a young man who later became a distinguished anthropologist at Cambridge University (Alfred Cort Haddon, 1855–1940). Those words are as true today as when they were written. Wrote Huxley:

I know of no department of natural science more likely to reward a man who goes into it thoroughly than anthropology. There is an immense deal to be done in the science pure and simple, and it is one of those branches of inquiry which brings one into contact with the great problems of humanity in every direction.

≈3≈

MAN'S NEAREST LIVING
RELATIVES

THE THEORY OF EVOLUTION postulates that organisms, plant and animal, have developed by a process of gradual continuous change from previously existing forms. In the years which have elapsed since Charles Darwin (1809–1882) proposed this idea in his epoch-making book *The Origin of Species* (1859), the theory of evolution has received an overwhelming amount of support from findings both in the field and in the laboratory. Darwin's theory rests on the principle of natural selection. This principle states (see p. 94) that those creatures are more likely to leave progeny who possess the traits which enable them to adapt themselves to the challenges of the environment, than those who do not possess such traits. Those variations are likely to be preserved, that is, selected, which are of adaptive value, and in this way evolutionary change is produced. The adaptively more fit are more likely to be effectively fertile than those who are adaptively less fit.

It should be noted that no mention is made here of "the survival of the fittest." That phrase, which became something of a shibboleth, is a holdover from the nineteenth century. It is an inaccurate description of adaptive fitness. "Darwinian fitness," as it is sometimes called, or adaptive fitness, implies generalized, *not* specialized fitness. Should the environmental challenges suddenly change, the fittest, the overspecialized, are likely to find themselves at a disadvantage. They will be unable to make the required adjustments rapidly enough, while those who have retained their generalized plasticity are much more likely to be able to do so. It is, therefore, not at all a matter of "the survival of the fittest," but very much a matter of the survival of the fit. In tracing the story of man's evolution this is very much a factor to be borne in mind.

The adaptively less fit will be outreproduced by the adaptively fit, and hence the attributes of the adaptively less fit will tend to disappear.

The mechanism of evolutionary change depends upon the inherent variability of the genetic constitution. The chemical packages, the DNA (deoxyribonucleic acid) molecules called genes, which are carried in the chromosomes contained in the cells of all animals, tend every so often to mutate in a random manner. A mutation is a permanent change in the structure of a gene resulting in a transmissible hereditary modification in the expression of a trait. Mutations are the raw materials of evolution. It is upon mutations that natural selection works, preserving only those which serve to render the organism adaptively more fit. In this manner changes in the genes can be transmitted to the offspring and produce new or altered attributes in the new individual. Since modifications by descent of physical traits produced in this way are the most immediately visible, the evolutionary relationships of plant and animal forms can be traced by the method of comparative morphology (*morphe* = form, *logos* = science), that is to say, by comparing their physical traits. In our own day evolutionary studies have been enlarged to include physiological, biochemical, serological, immunological, and genetic traits. However, the study of evolutionary relationships remains essentially comparative.

Some wit once remarked that man is descended from the apes, and has been descending ever since! Neither statement is quite true. It is a common belief that man is descended from "the monkeys," meaning the kinds of monkeys living today, but no scientist has ever held this belief. Nor is it true that man is descended from any of the living apes. What, then, is the relationship of man to the monkeys and apes? Is man in any way connected with these creatures? How does one go about finding out whether he is or not?

PUTTING MAN IN HIS PROPER PLACE

The best way to find out in which group any animal belongs is to ask the question: What kind of animal does our specimen most closely resemble? In our case the specimen is man. Does he resemble a fish? Not really. A cat? No. A bird? Most certainly not. A snake? Of course not. An elephant? No, no. A monkey? Well, yes, he resembles a monkey a great deal more than he does any of the other animals we have mentioned.

Well, what monkey does he most resemble? The fact is that he most resembles an ape rather than a monkey. As Oliver Goldsmith put it,

Of Beasts it is confessed the ape
Comes nearest us in human shape.

There are four apes: the gibbon, the orangutan, the chimpanzee, and the gorilla. Which does man most resemble? Most scientists say the gorilla. Very well, then. Man most closely resembles the gorilla of all the living creatures on this earth. He also bears striking resemblances to the monkeys. The monkeys bear striking resemblances to the little tarsiers of Malaysia, and these bear significant resemblances to the lemurs of Madagascar and the Comoro Islands.

The lemurs, tarsiers, monkeys, apes, and men seem, then, to form a common group in virtue of their physical resemblances to each other. This they do, and it was precisely for this reason that Linnaeus put them all into the same large group of animals, the Order of Primates. (An *order* is a classificatory group within a larger *class* of animals that bear certain structural resemblances to each other.) "Primate" means "first in rank or order." By the name "primates," Linnaeus meant to suggest that this was the foremost or highest order of animals.

We owe the classificatory method of dealing with animals and plants to Linnaeus, who published it in his *Systema Naturae* in 1735, the 1758 edition of which serves as the basis of Linnaean nomenclature. The point and the purpose of all classification is to provide a simple, practical means by which students may know what they are talking about and enable others to find out. In Table 1 will be found a classification of the primates with the subdivisions, scientific names, and popular names.

It should be noted that the names given to suborders and to superfamilies often terminate in "oidea," while the names given to families terminate in "idae," and those given to subfamilies in "inae." In the Linnaean classification the genera are given as proper names without any categorical stem ending. The name of

TABLE 1. CLASSIFICATION OF THE PRIMATES

Suborder	Infraorder and Series	Superfamily	Family
PROSIMII	LEMURIFORMES	LEMUROIDEA	Lemuridae
			Indridae
			Daubentonioidea
	LORISIFORMES	LORISOIDEA	Lorisidae
			Galagidae
	TARSIIFORMES	TARSIOIDEA	Tarsiidae
ANTHROPOIDEA	PLATYRRHINI	CEBOIDEA	Callithricidae
			Cebidae
	CATARRHINI	CERCOPITHECOIDEA	Cercopithecidae
		HOMINOIDEA	Hylobatidae
			Pongidae
			Hominidae

Subfamily	Genus	Common Name
Lemurinae	*Lemur*	True Lemurs
	Hapalemur	Gentle Lemurs
	Lepilemur	Sportive Lemurs
Cheirogaleinae	*Cheirogaleus*	Dwarf Lemurs
	Microcebus	Mouse Lemurs
	Phaner	Fork-crowned Dwarf Lemurs
	Indris	Endrina
	Propithecus	Sifaka
	Lichanotus	Woolly Avahi
Daubentoniidae	*Daubentonia*	Aye-Aye
Lorisinae	*Loris*	Slender Loris
	Nycticebus	Slow Loris
	Arctocebus	Angwantibo
	Perodicticus	Potto
Galaginae	*Galago*	Typical Bush Babies
	Galagoides	Dwarf Bush Babies
	Euoticus	Needle-clawed Bush Babies
	Tarsius	Tarsier
Callithricinae	*Callithrix*	Tufted Marmoset
	Cebuella	Pygmy Marmoset
	Mico	Naked-eared Marmoset
	Marikina	Bald-headed Tamarin
	Tamarin	Black-faced Tamarin
	Tamarinus	Mustached Tamarin
	Leontocebus	Maned Tamarin
	Oedipomidas	Pinché
Callimiconinae	*Callimico*	Goeldi's Monkey
Aotinae	*Aotes*	Night Monkey
	Callicebus	Titi Monkey
Pitheciinae	*Pithecia*	Hairy Saki Monkey
	Chiropotes	Short-haired Saki
	Cacajao	Ouakari Monkey
Cebinae	*Saimiri*	Squirrel Monkey
	Cebus	Capuchin Monkey
Atelinae	*Lagothrix*	Woolly Monkey
	Ateles	Spider Monkey
	Brachyteles	Woolly Spider Monkey
Alouattinae	*Alouatta*	Howler Monkey
Cercopithecinae	*Cercopithecus*	Guenons
	Erythrocebus	Red-haired Patas
	Cercocebus	Mangabey
	Macaca	Macaque
	Cynopithecus	Celebes or Black Ape
	Theropithecus	Gelada Baboon
	Papio	Baboon
	Mandrillus	Mandrill, Drill
Colobinae	*Presbytis*	Common Langur
	Colobus	Guereza
	Rhinopithecus	Snub-nosed Langur
	Pygathrix	Douc Langur
	Simias	Pagi Island Langur
	Nasalis	Proboscis Monkey
Hylobatinae	*Hylobates*	Gibbon
	Symphalangus	Siamang
Ponginae	*Pongo*	Orangutan
	Pan	Chimpanzee
	Gorilla	Gorilla
	Homo	Man

the species always follows that of the genus, and that of the subspecies follows that of the species. The name of the genus always begins with a capital letter, while the names of species and subspecies begin with a small letter, and each of these names is *italicized* (as in this latter word). For example, *Homo sapiens neanderthalensis* means the form of man like ourselves belonging to the Neanderthal variety.

The primates form a distinct and recognizable group. So far so good. But from what and where did the primates originate?

THE ORIGIN OF THE PRIMATES

The creatures that most look like lemurs—the most primitive members of the Primate Order—are known as insectivores or oriental tree shrews. They are to be found mostly in the Malaysian region, and also in China and India. The group of tree shrews that look most like the lemurs are known as tupaias. The tupaia looks something like a long-snouted mouse, and is actually not much bigger. The earliest insectivores are known from the Cretaceous geological epoch, which goes back to more than 75 million years ago. That is when it is believed the first pre- or proto-primates may have come into being, somewhere in the early Paleocene epoch. (Examine, and try to memorize, Table 2, which lists the geological eras, periods, and epochs, and the forms of life typically associated with them.) Some authorities in the past have considered that the tupaias, the earliest fossil forms of which are known from the early Oligocene epoch, should be classed with the primates, but other authorities today believe that the resemblances are superficial, and that the tupaias properly belong with the insectivores.

The earliest primates probably resembled tree shrews, and like them may have been predominantly insect-eaters, who lived mostly among the lower branches of bushes and trees. Their brains would have been largely olfactory (smell) brains. Since these early primates earned their living mainly by keeping their noses to the ground, as it were, by smelling out their food, their brains would have been largely organized along olfactory lines. Their eyes must have been situated on each side of the face rather than in front (as in ourselves), and their tails probably helped them to balance themselves on the limbs of trees. Like

Tree shrew (*Urogale
everetti*), Philippines

tree shrews they were almost certainly nocturnal animals, that
is, animals that sleep during the daylight hours and are wakeful
only during the hours of the night.

THE LEMURS: FAMILY LEMURIDAE
The lemurs are foxlike creatures that vary in size from the size
of a mouse to an average-sized dog. The small lemurs are
nocturnal, the large ones diurnal (awake during daylight and
asleep at night). They have nonprehensile (prehensile = grasp-
ing) tails which they use for balancing, and the eyes tend to be
more toward the front of the face than in the tupaias. They
seem to divide their time between a life on the ground and a life
in the trees. The olfactory part of the brain is somewhat smaller
in size, while the back part of the brain associated with vision is
more highly developed than in the tupaias. The brain is, however, still to a significant extent an olfactory one, and in the
parts of the world where lemurs live, Madagascar, Malaysia,
Africa, Ceylon, and southern India, one may see lemurs of

TABLE 2. GEOLOGICAL TIME SCALE OF THE APPEARANCE OF VARIOUS REPRESENTATIVE FORMS OF LIFE

Era	Period	Epoch	Millions of Years Since the Beginning of Each Epoch
CENOZOIC The Age of Mammals	Quaternary	Recent	1/40
		Pleistocene	2
		Pliocene	12
	Tertiary	Miocene	28
		Oligocene	39
		Eocene	58
		Paleocene	75
MESOZOIC The Age of Reptiles	Secondary	Cretaceous	135
		Jurassic	165
		Triassic	205
PALEOZOIC The Age of Ancient Life		Permian	230
		Carboniferous	255
		Devonian	325
		Silurian	360
		Ordovician	425
		Cambrian	505
PROTEROZOIC			925
ARCHEOZOIC			1,500

The estimated number of millions of years in the fourth column for the Tertiary period is based upon a combination of paleontological data, with specific reference to the evolution of the horse from *Hyracotherium* to *Equus,* and the evidence of geology and radioactivity. The figures for the preceding periods are largely based on the uranium transformation method. When uranium and lead occur together in a fragment of rock otherwise free from these elements, it may generally be safely assumed

Forms of Life

Man, the slave and master.

Australopithecines, *Homo erectus, Swanscombe, Homo sapiens.*

Kenyapithecus. First men probably appeared during the latter part of this epoch.

Appearance of true anthropoid apes. *Dryopithecus, Sivapithecus, Proconsul.*

Primitive anthropoid apes appear such as *Propliopithecus.*

Spread of modern mammals. Tarsiers.

Appearance of insectivorous preprimates and earliest primates, primitive lemuroids and tarsioids.

Rise of archaic mammals and birds. Extinction of dinosaurs, pterodactyls, and toothed birds. Insectivores.

Spread of primitive mammals and pterodactyls, rise of toothed birds.

Rise of dinosaurs, pterodactyls, and primitive mammals.

Spread of amphibians and insects. Extinction of trilobites.

Primitive reptiles, insects, spiders. Great forests of ferns and mosses.

Rise of fishes and amphibians. Spreading of forests.

Rise of ostracoderms, sea-scorpions (Eurypterids). First land plants.

First primitive fishes, the ostracoderms.

Still no land-life known, trilobites, mollusks, brachiopods.

Sponges, protozoons, diatoms, and protophyta, and other commencing complex forms of life developed during this era.

Probably simple unicellular sea-dwelling forms.

that the lead represents "decomposed" or transformed uranium. It is known that 1,000,000 grams of uranium yields 1/7600 grams of lead a year. Hence the age of such rocks can be determined from the proportions of these elements which they contain, thus:

$$\text{Age of rock} = \frac{\text{Weight of lead}}{\text{Weight of uranium}} \times 7600 \text{ million years}$$

Ring-tailed lemur (*Lemur catta*), Madagascar

various shapes and forms still relying mostly upon the sense of smell in order to gain a living. The lemurs appear to have gone on a merry spree of specializing in many different forms, and while they have managed to exist for a very long time, they don't seem to have given rise to any more developed or advanced types.

THE TARSIERS: GENUS *Tarsius*
The tarsiers live in the Malay Archipelago and the Philippines. They are about eight inches long, with long, tufted tails, flat faces, and the most enormous eyes, set in the front of their faces, of any of the primates. This animal is often called the spectral tarsier, "spectral" because of its large eyes and "tarsier"

because of the greatly elongated tarsal bones in each foot. (The heel bone, or calcaneus, and the navicular at its side together form part of the tarsus.) The tarsier is nocturnal, and spends most of his time in the trees. *Tarsius* possesses the peculiar capacity of being able to swivel his head around an orbit which completes 180 degrees so that his nose is in line with his spinal column, a feat which has been celebrated in the limerick:

> The Tarsier, weird little beast
> Can't swivel his eyes in the least,
> But when sitting at rest
> With his tummy due west
> He can screw his head round to face east.

A very interesting thing about the tarsiers is exhibited in the smallness of the olfactory parts of the brain as compared with the same parts in the lemurs. In correlation with this the snout shows reduction, but the eyes have enlarged. Notice this association, because it is most significant. Tarsiers are really arboreal creatures, that is to say, they are adapted for living in trees. A monkey in the trees has to gain its living in a way quite different from one who lives on the ground. On the ground the animal needs its sense of smell most. In the trees the sense of smell isn't of much help; what the animal needs most is vision, good vision. In order to catch flies and other insects, it must be able to follow their movements and those of any other small creatures that may be wandering around in the trees. And in the trees it also needs to free its forelimbs so that it can catch and deal with these creatures with its hands. This the tarsiers have done. They have well-developed hands, but the thumbs are not truly opposable, that is, like thumbs that can turn on their own axes to face the palmar surface of the fingers.

Most authorities believe that from the tarsiers the monkeys and apes arose. The first tarsiers are found in the Middle Paleocene epoch of North America, and because the earliest tarsioids (tarsiuslike creatures) and lemuroids are found in North America, it is this continent which is believed to be the birthplace of the primates.

The tarsiers are important for us to study because they show the beginning development of what probably constitutes the most important single factor in the evolution of the primates,

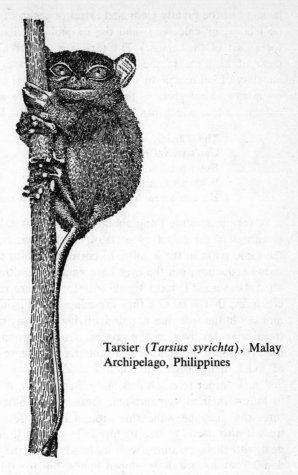

Tarsier (*Tarsius syrichta*), Malay
Archipelago, Philippines

namely, the increasing importance and dominance of the sense
of vision and the correlated changes associated with it. This
vastly increased what has been called "the space of recognition." In other words, it greatly enlarged the possibilities of
experience for the primates.

Like most of the other nocturnal primates, *Tarsius* lives in
small family groups rather than in large communities, though
there is, of course, some sort of cross-communication between
families or members of families within the larger population of
tarsiers. Those who have managed to make pets of them find
them to be very endearing little creatures.

Spider monkey (*Ateles ater*),
Mexico to Amazon basin

THE SOUTH AMERICAN MONKEYS OR NEW WORLD
MONKEYS: FAMILIES CALLITHRICIDAE AND CEBIDAE
The South American monkeys range in size from the tiny
marmosets to the *Ateles* monkeys (who have no thumbs but
highly prehensile tails). *Ateles* often attains a height of almost
three feet. The durukulis or night monkeys of Nicaragua, Peru,
and the Amazon constitute good links between the tarsiers and
the monkeys, for the durukulis are nocturnal, small in size, and
live in small family groups. Otherwise they are much more like
monkeys than they are like tarsiers.

All South American monkeys are arboreal. Less than half of

Barbary ape (*Macaca sylvana*),
Gibraltar, Asia, Malaysia, Africa

the genera have prehensile tails, and most of them don't have opposable thumbs. What they lack in opposability of thumbs many of them make up with prehensility of tails. The *Ateles* or spider monkeys can actually use their tail as a sort of third hand, being able to pick up all sorts of things with it.

The brain of the South American monkey is pre-eminently a seeing one, with the functions and structures subserving the sense of smell very considerably reduced. Though it is questioned by some, it is quite probable that the South American monkeys gave rise to the Old World or Eastern Hemisphere monkeys. However, the possibility remains that each family may have developed independently from a common ancestral stock.

THE EASTERN HEMISPHERE MONKEYS OR
OLD WORLD MONKEYS: FAMILY CERCOPITHECIDAE
All the Old World monkeys are Asiatic and African in distribution. The exception is the European Barbary ape of Gibraltar.
The Old World monkeys are very like their New World cousins
except that they do not have prehensile tails, their thumbs are
opposable, and they have only two premolar or bicuspid teeth
(the teeth immediately behind the eye or canine teeth) on each
side of the jaw instead of the three that the New World monkeys
have. There are all sorts of Old World monkeys, some of them
reaching the size of big dogs, such as the baboons, some of
whom are fetchingly colored with scarlet and blue faces and
often equally garishly colored seats. Not all baboons are so
flamboyantly decorated, only, it appears, the most doglike of
them.

The brains of the Old World monkeys are distinctly seeing
ones; smell is poorly developed, and intelligence does not seem
to be any more highly developed in them than it is in the New
World monkeys. In fact, there is a most remarkable identity in
form and appearance of the brain in the two great groups of
monkeys. This probably represents the persistence, over the
course of some 50 million years, of a complex trait that was,
and still is, characteristic of the original New World monkeys,
and which has endured over this immensely long period of time
in the Old World monkeys.

The Old World monkeys undoubtedly gave rise to some apes,
all of which became extinct long ago. Many of these are known
from their fossil remains, enabling us to reconstruct something
of the evolution of the primates. The early tarsiers may very
well have been the ancestors of early intermediate forms before
the apes finally arose from these. The intermediate forms were
either very like Old World monkeys or identical with them.

THE APES: FAMILIES HYLOBATIDAE
AND PONGIDAE

The apes consist of the gibbon, orangutan, chimpanzee, and
gorilla.

THE GIBBON: *Hylobates* and *Symphalangus*

The gibbon (genus *Hylobates*) is the acrobat among the primates, for he has developed the most amazing agility in the trees. Here he can negotiate his way faster than a man can run on the ground. When he stands erect, his fingers touch the ground, so long are his upper extremities. He is an inhabitant mainly of Asia. Another form of gibbon, the siamang (genus *Symphalangus*) is found on the island of Sumatra. Ever since the Oligocene epoch, some 39 million years ago, the gibbons seem to have pursued their independent evolution, quite apart from any of the other apes. Fossil forms of the gibbon from that epoch indicate that there has not actually been much change over this long period of time. The gibbons rarely reach three feet in height. They have a small brain volume, about 100 cubic centimeters (cc), but a highly developed visual capacity. They certainly stand a long way off from the direct line of man's ancestry.

THE ORANGUTAN: *Pongo*

The orangutan (Malay for "wild-man-of-the-woods"; genus *Pongo*) is a delightful creature who lives exclusively in Borneo and Sumatra. He is entirely arboreal, with shaggy, reddish-brown hair and a face that is often pathetically manlike. He has a good-sized brain (about 400 cc) with intelligence to match, and is reputedly the strongest creature in the jungle. Orangs have been known, when attacked, to pull the jaws of giant crocodiles apart and then fling the victim into the water as if it were a broken reed. Yet the male orang weighs no more than 165 pounds, while the female rarely weighs more than 80 pounds.

THE CHIMPANZEE: *Pan*

More manlike than the orangutan is the chimpanzee (genus *Pan*). The chimpanzee is an inhabitant of western and central equatorial Africa. The male weighs about 110 pounds and the female about 88 pounds. The average height of the male is five feet and of the female four feet. Chimpanzees are expert tree climbers, but they spend most of their time on the ground. They have a brain volume of about 400 cc. They are quite intelligent

Chimpanzee (*Pan troglodytes*), western and central equatorial Africa

creatures and, as everyone knows, can be taught many tricks requiring great control and intelligence.

Chimpanzees do not have a bridge, or raised bones, to their noses, their jaws are rather projecting, and they have quite well-developed canine teeth. Like the orang they are predominantly vegetarian, and among themselves and with all other creatures are thoroughly peaceful.

It should be remembered that when any of the apes, or any other animals, are seen in a zoo, they are in the most unnatural of conditions. Quite frequently they are isolated from their kind, but even when this is not so, they are usually most unhappy and only too frequently mentally ill—just as we would be if we were treated, or rather mistreated, in a similar manner. Under conditions of confinement, the chimpanzee seems to become manic

TABLE 3. RESEMBLANCES AND DIFFERENCES BETWEEN
THE VARIOUS KINDS OF PRIMATES

Trait or Character	Lemurs	Tarsiers	New World Monkeys Marmosets	Ceboids
Habitat	Madagascar, Africa and Asia	Malay Archipelago, Philippines	Mexico and South America	South America
Number of Genera	18	1	8	11
Number of Species	34	3	35	39
Number of Subspecies	100	12	51	161
First Sexual Skin (in years)	None	*	Present	None
First Menstruation (in years)	Absent	Present	Non-menstrual bleeding	Present
Duration of Menstruation (in days)	Absent	Absent	*	1-5
Duration of Menstrual Cycle (in days)	*	26	*	*
Time of Ovulation (counting from first day of menses)	*	*	*	*
Breeding Season	April	Unrestricted	· Unrestricted	Unrestricted
Interval Between First Menstruation and First Conception	*	*	*	*
Duration of Pregnancy (in days)	125	*	145	139
Duration of Labor (in hours)	*	*	*	*
Birth Weight (in percentage of adult female weight)	4.6	*	*	7.0-8.5
Care of infant	Mother	Mother	Father	Mother usually
Suckling (in months)	6+	Several	Several	Several
Multiple Births	Less frequent than single	Rarely	Usual	Occasionally
Completion of Permanent Dentition	*	*	*	6 years
Life Span (in years)	14	*	10	14
Brain Weight (in grams)	30	3.6	10	45-112
Brain Weight/Body Weight Ratio	13.5	39	39	10-70

* Data not available

Old World Monkeys	Gibbon	Orangutan	Chimpanzee	Gorilla	Man
Africa and Asia	Asia and Malaysia	Borneo and Sumatra	Central Africa	Central Africa	Global
12	2	1	1	1	1
88	7	2	1	1	1
245	22	0	2	2	0
2 years 6 months	None	Present in some during pregnancy	8 years	Absent	Absent
2 years 10 months	8 years 6 months	Present	8 years 11 months	9 years 6 months	13½
5	2½	*	4	4	5
27	30	29	35	30	28
13	*	*	14	14	14
Unrestricted	Unrestricted	Unrestricted	Unrestricted	Unrestricted	Unrestricted
1 year 6 months	*	*	1 year 3 months	1 year 6 months	3 years
170	210	275	231	259	266½
3½	*	*	5	*	First child 14 Later children 8
4.6-7.0	7.5	4.1	4.0	2.4	5.5
Mother	Mother	Mother	Mother	Mother	Mother
6+	6+	36+	24+	36+	12+
About 1.05%	Rarely	Very rare	6% in captivity	Very rare	1.1%
6 years 10 months	9 years	9 years 8 months	10 years 2 months	10 years 6 months	19 years 9 months
25	30	30	35	35	75
60-200	96	371	345	425	1320
12-30	21	5.4	12	4.7	28

TABLE 3 (*continued*)

Trait or Character	Lemurs	Tarsiers	New World Monkeys Marmosets	Ceboids
Brain Volume (in cubic centimeters)	*	*	*	45-112
Brain Surface	Unconvoluted	Unconvoluted	Unconvoluted	Richly convoluted
Stereoscopic Vision	Absent	Doubtful	Some	Present
Color Vision	Doubtful	Doubtful	Doubtful	Present
Habitus and Waking Activity	Mostly nocturnal	Nocturnal	Diurnal	Mostly diurnal
Mode of Life	Mainly arboreal	Arboreal and terrestrial	Arboreal	Arboreal
Social Aggregates	Family groups	Family groups	Family groups	Small communities
Shelter	In trees	At base of bushes	In trees	In trees
Diet	Frugivorous	Insectivorous	Frugivorous and insectivorous	Mainly frugivorous
Locomotion	Quadrupedal	Leaping	Climbing	Climbing
Height or Length	4-24 inches	8 inches	10 inches	10-30 inches
Weight (in pounds)	5	½	½	20
Dental Formula	I$\frac{2}{2}$, C$\frac{1}{1}$, PM$\frac{3}{3}$, M$\frac{3}{3}$	I$\frac{2}{1}$, C$\frac{1}{1}$, PM$\frac{3}{3}$, M$\frac{3}{3}$	I$\frac{2}{2}$, C$\frac{1}{1}$, PM$\frac{3}{3}$, M$\frac{2}{2}$	I$\frac{2}{2}$, C$\frac{1}{1}$, PM$\frac{3}{3}$, M$\frac{3}{3}$
Length of Hindlimbs Compared to Forelimbs	Hindlimbs longer than forelimbs	Hindlimbs longer than forelimbs	Hindlimbs much longer	Hindlimbs usually longer
Thumb Length in Percentage Proportion to Hand Length	46	*	50	50
Thumb	Opposable	Opposable	Not opposable	Not opposable
Big Toe	Opposable	Opposable	Opposable	Opposable
Nails or Claws	Nails except on 2nd toe	Nails except on 2nd and 3rd toes	Claws except big toe	Nails
Tympanic External Auditory Meatus	Absent	Present	Present	Absent

* Data not available

Old World Monkeys	Gibbon	Orangutan	Chimpanzee	Gorilla	Man
-200	97.5	416	400	550	1350
Richly convoluted	Richly convoluted	Richly convoluted	Richly convoluted	Richly convoluted	Richly convoluted
Present	Present	Present	Present	Present	Present
Present	Present	Present	Present	Present	Present
Diurnal	Diurnal	Diurnal	Diurnal	Diurnal	Diurnal
Mainly arboreal	Arboreal	Arboreal	Mainly terrestrial	Terrestrial	Terrestrial
Small communities	Small bands	Family groups	Small bands	Small bands	Large groups
Mostly in trees	In trees	Nests in trees	Lower branches of trees	Lower branches of trees	In roofed buildings
Mainly frugivorous	Vegetarian insects, and small birds	Frugivorous	Frugivorous	Herbivorous	Omnivorous
Climbing	Brachiation	Brachiation	Obliquely quadrupedal	Obliquely quadrupedal	Bipedal
2-35 inches	3 feet	4 feet	5 feet	5 feet 6 inches	5 feet 6 inches
	13	♂ 165, ♀ 81	♂ 110, ♀ 88	♂ 375, ♀ 200	♂ 150, ♀ 130
$I\frac{2}{2}, C\frac{1}{1}, PM\frac{2}{2}, M\frac{3}{3}$	$I\frac{2}{2}, C\frac{1}{1}, PM\frac{2}{2}, M\frac{3}{3}$	$I\frac{2}{2}, C\frac{1}{1}, PM\frac{2}{2}, M\frac{3}{3}$	$I\frac{2}{2}, C\frac{1}{1}, PM\frac{2}{2}, M\frac{3}{3}$	$I\frac{2}{2}, C\frac{1}{1}, PM\frac{2}{2}, M\frac{3}{3}$	$I\frac{2}{2}, C\frac{1}{1}, PM\frac{2}{2}, M\frac{3}{3}$
Hindlimbs longer than forelimbs	Forelimbs much longer	Forelimbs much longer	Forelimbs longer	Forelimbs longer	Hindlimbs longer
5	54	44	47	50	68
Opposable	Opposable	Opposable	Opposable	Opposable	Opposable
Opposable	Opposable	Opposable	Opposable	Opposable	Not opposable
Nails	Nails	Nails	Nails	Nails	Nails
Present	Present	Present	Present	Present	Present

TABLE 3 (*continued*)

Trait or Character	Lemurs	Tarsiers	New World Monkeys Marmosets	Ceboids
Nostrils	Widely separated	Widely separated	Widely separated	Widely separated
Laryngeal Sacs	Absent	Absent	Absent	Absent
Pterion	Parieto-sphenoidal	Protomalar parieto-sphenoidal	Malarparieto-sphenoidal	Malarparieto-sphenoidal
Lacrimal Foramen	External	External	Internal	Internal
Entepicondylar Foramen	Present	Present	Present	Present
Chromosome Number	44-60	80	46	44-54
Ischial Callosities	Absent	Absent	Absent	Absent
Tail	Non-prehensile	Non-prehensile	Non-prehensile	Prehensile in half the genera

and hyperactive, whereas the gorilla grows depressed and despondent.

The chimpanzee does not walk erect, although he can do so when he desires; his habitual mode of progression is in an obliquely quadrupedal position with the knuckles of his digits bent to support him in front. He has a short opposable thumb, between which and his forefinger he can hold and thread a needle with the best of needleworkers.

Dr. Ludwig Heck, the head of the Zoological Gardens in Berlin, who has had much experience with chimpanzees, considers that they are kind and generous by nature, and he describes how they will give away part of their own food to a hungry companion and go to one another's aid when such help is needed. One of his chimpanzees succeeded in pulling a splinter out of a keeper's hand, and made a very neat job of it, too.

All authorities are agreed that the modern chimpanzee does not stand in the direct line of man's ancestry. The chimpanzee, in common with the orang and gorilla, probably originated during the Miocene epoch some 28 million years ago, or at least

Old World Monkeys	Gibbon	Orangutan	Chimpanzee	Gorilla	Man
Narrowly separated	Narrowly separated	Narrowly separated	Narrowly separated	Narrowly separated	Narrowly separated
Absent	Mostly in *Symphalangus*	Present	Present	Present	Absent
Mostly fronto-temporal	Spheno-parietal	Mostly spheno-parietal	Fronto-temporal	Fronto-temporal	Spheno-parietal
Internal	Internal	Internal	Internal	Internal	Internal
Absent	Absent	Absent	Absent	Absent	Absent
2-72	44	48	48	48	46
Present	Present	In 5%	In 38%	7%	Absent
Non-prehensile	Absent	Absent	Absent	Absent	Absent

their immediate ancestors did. We have the fossil remains from this period of chimpanzeelike creatures known as the subfamily Dryopithecinae or the genus *Dryopithecus* from India and from Europe.

Ever since the Miocene epoch the lines which led to the chimpanzee and the gorilla (and the orang) have been pursuing their own evolutionary destiny quite apart from man.

THE GORILLA: GENUS *Gorilla*
There are two kinds of gorilla, which differ very slightly from one another, the lowland or coastal gorilla of western equatorial Africa, known as *Gorilla gorilla gorilla* (yes three times, giving the genus, species, and subspecies or race), and the east central African highland or mountain gorilla, *Gorilla gorilla beringei* (after Captain von Beringe, a German officer who first described it). Because of their close resemblance there is a growing tendency among taxonomists to place the gorilla with the chimpanzee in the same genus, *Pan*.

The gorilla is the bulkiest of the primates. Gorillas are not as tall as most people imagine, being on the average only about

Male gorilla (believed to
be *Gorilla gorilla beringei,*
the mountain gorilla),
western and eastern
equatorial Africa

five and a half feet. Weights up to 670 pounds have been recorded. As in the other apes, the forelimbs are longer than the
hind limbs. Undoubtedly the gorilla is the strongest of all
primates. Gorillas have bare chests, not hairy ones as is commonly believed. Otherwise, they are covered with black hair
except on the face. Some of the hairs on the top of the head
may be variously colored. While the orang can grow a beard
and mustache, the chimpanzee and gorilla cannot.

The average male cranial capacity is 550 cc, while that of the
female is about 460 cc. Males sometimes attain a cranial
capacity of as much as 750 cc. The face is most interesting
because the nasal bones show a slight elevation, so that the

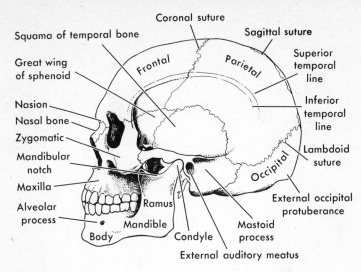

The human skull

nose, while still quite flattish, is more like that of man than the nose of any of the other apes. The lips, as in all the apes, are thin with hardly any of the reddish-looking mucous membrane furled outward as in ourselves.

Gorillas are terrestrial creatures, that is, they live on the ground. But they are good tree climbers. They build nests, mostly on the ground, in which to sleep, and never use these more than once. Their diet is principally herbivorous, that is, they feed chiefly on plants. Gorillas usually live in small bands of from three to ten or more families. During the day they generally separate but return before or at nightfall to make their nests together. All who have observed gorillas in their native habitat describe them as the most amiable of creatures. All the apes are peaceful animals, and with the exception of the chimpanzee who will occasionally kill and eat a monkey, they will never molest any living thing unless seriously provoked or badly frightened.

The gorilla moves about in much the same way as the chimpanzee, in an obliquely quadrupedal position, and like the chimpanzee the gorilla can move, when it wants to, in the erect position.

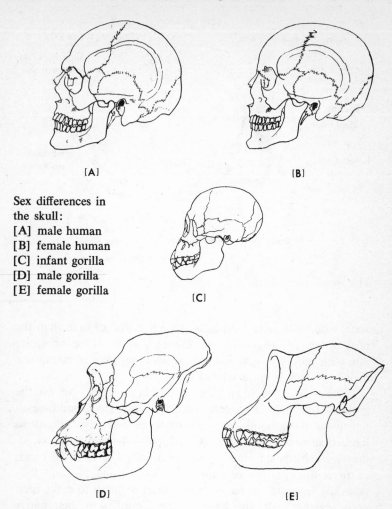

Sex differences in
the skull:
[A] male human
[B] female human
[C] infant gorilla
[D] male gorilla
[E] female gorilla

[A]

[B]

[C]

[D]

[E]

THE APES SUMMARIZED

All apes are without tails whereas all monkeys have tails. All
apes have both opposable thumbs and big toes, and all of them
have flattened nails on their fingers and toes. Some of the lower
monkeys have claws. The great apes (this excludes the gibbon)
are all over four feet in height, have largish brains, and are the
most intelligent of the nonhuman primates. They are all diurnal
in their habits, mostly vegetarian, and live in small family

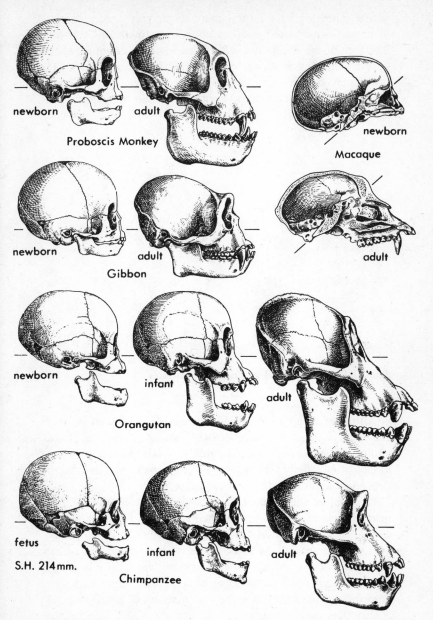

newborn

adult

Proboscis Monkey

newborn

Macaque

adult

newborn

adult

Gibbon

newborn

infant

adult

Orangutan

fetus

S.H. 214 mm.

infant

adult

Chimpanzee

Age changes in catarrhine skulls
(After A. H. Schultz)

bands. They all have large canine teeth, very unlike those of men, and are covered with an abundant growth of body hair. The face of the gorilla is always black, but the chimpanzee will vary all the way from white to mottled brown and black. The orang may be white-skinned or mottled-brown. The gibbon is usually black-skinned, though his body hair may vary from white to black.

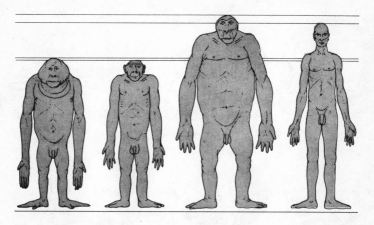

Apes and man compared (After A. H. Schultz)

All the great apes are capable of making tools, that is, artificial implements, by modifying twigs, sticks, straw, and leaves, and of using them as tools, but that is about the extent of their toolmaking activity. They are not, as man is, continuously dependent upon tools for their existence.

In structure the apes living today are far too different from men to be in the direct line of their ancestry. They are at most remote collateral relatives or distant "cousins" of ours. Today's apes are too specialized to be ancestral to man, that is to say, they have developed all sorts of special traits, like large eye (canine) teeth, and in the case of the gorilla great bulk. The immediate nonhuman ancestors of man must have been less apelike than the great apes.

Let us see in the next chapter what is known of the possible ancestors of man.

ᚦ4ᚦ

THE ANCESTRY
OF MAN

MAN'S PROBABLE ANCESTORS

THE IMMEDIATE ANCESTORS of the earliest men were most likely not so specialized, that is, had not developed any irreversible special adaptations to a particular habitat or mode of life, as the great apes, and probably retained many primitive traits, in many respects resembling men more than they did apes. Such creatures would have preserved a strong tendency not to develop tusklike eye teeth, great weight, sagittal crests on top of the skull for the attachment of the temporal muscles that move the lower jaw up and down, or other such specializations, but would have maintained a somewhat conservative tendency to enlarge on the endowments they already possessed. For example, the tendency in the evolution of the primates has been for brain size to increase. Any primate group evolving in the direction of increase in brain size, in adaptation to the challenges of the environment, especially through artificially developed responses, would in the long run stand a good chance of evolving traits which would lead to the achievement of human status. Those individuals able to use their brains effectively in response to the challenges of the environment, who were able to make the most appropriately successful responses to the challenges of the environment, would be more likely to perpetuate their kind, than those who failed to make such responses.

Such creatures would first have developed the upright posture, in adaptation to the requirements of hunting. Second, because of their increasing dependence on tools, they would never have developed the large, overspecialized canine teeth because they would have been unnecessary for omnivores; thus they would not have needed so much bone in the upper and lower jaws for them, and the snout region would have been reduced in the direction of the rather straight-facedness of man. Third, the brain would have been enlarged because of the necessity of a large enough warehouse in which to store the required information, with corresponding changes in the form of the skull, in the direction of a more domelike top of the head and the development of a forehead. Rearrangement of the facial bones and the rendering more vertical of the front of the head would produce an elevation and forward projection of the nasal bones—which in the monkeys remain flat—culminating in the unique structure, the peninsula of bone, cartilage, and soft tissues, which we know as the nose of man.

It is quite probable that man's immediate ancestors as a group had already lost a considerable part of the hairy coat which is characteristic of all anthropoids, indeed, of all primates. What the skin color was we cannot be sure, but since the ancestors of man were almost certainly tropical animals of African origin, they may have been black-skinned.

Did they have speech? We don't know. Perhaps they had the rudiments. Did they make tools? It is probable that they did. In various parts of Africa pebble tools are found in great numbers. These are now attributed to the australopithecines.

THE EARLIEST MEN—THE AUSTRALOPITHECINES?

In South and East Africa there have been unearthed a large number of skeletons of fossil apes, which have been named the australopithecines (*australis* = south, *pithecus* = ape). The remarkable thing about the australopithecines is that they are in most respects apelike, except that their brain capacity is larger and that their hipbones and the bones of the thigh (femur), leg (tibia and fibula), and foot are manlike.

From the structure of the hip and leg bones, it is quite evident that the australopithecines walked either erectly or almost so. Thus now, for the first time, we have clear evidence of the order of the functional evolution of some of the parts of the human body. The erect posture was attained before the brain evolved to a large size. Some authorities used to think that it was the other way around. Now we know with certainty that man's ancestors stood erect first, before their brains increased in size.

How long ago did the australopithecines live? We don't know with certainty because the geology (the study of the earth) of the regions of Africa in which the australopithecine remains are found is not well understood. But most authorities are of the opinion that these fossil remains date back to the Plio-Pleistocene boundary, that is to say, well over 2 million years ago. They may have died out in Africa about a quarter of a million years ago or even later, but that doesn't mean that they could not have been ancestral to man or closely related to the group of ancient apes which were the direct ancestors of man.

What may well have been an australopithecine was discovered in 1965 in early Pleistocene sediments in the Kanapoi drainage in southeastern Turkana (northwestern Kenya). This find consisted of the lower end of a left humerus which is strikingly manlike in form and readily distinguishable from the same bone in the chimpanzee and gorilla. This fossil has been named Kanapoi Hominid I. The sediments in which the bone was found yielded a potassium-argon date of 2.5 million years. No artifacts of any kind were found.

The australopithecines had an average brain size of less than 600 cc; the average brain size of living men today is about 1350 cc. It might therefore be thought that the australopithecines would have had quite a long way to go before they attained the status of man. They would have had to add approximately an additional 400 cc to their brain volume to reach the brain size of *Homo erectus erectus*. A rapid jump of 400 cc is inconceivable.

Furthermore, the skull form of all australopithecines is extremely apelike. The vault of the skull is low, and in some forms like *Australopithecus* (*Paranthropus*) *crassidens* found at Kromdraai, South Africa, a well-developed crest, similar to that of the male gorilla, runs from back to front on the top of the head. (This sagittal crest, as it is called, serves for the attachment of the massive muscles at the sides of the head which move the lower jaw up and down.) In others, such as *Aus-*

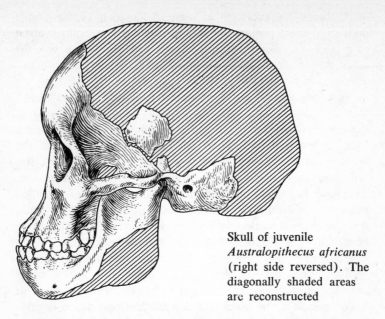

Skull of juvenile
Australopithecus africanus
(right side reversed). The
diagonally shaded areas
are reconstructed

tralopithecus robustus found at Swartkrans, South Africa, the lower jaw is of massive thickness. But the fact is that the australopithecines do exhibit many manlike traits, which without any doubt place them within the genus of man. The teeth, for example, are more like those of human beings than those of any other known creature. So are the bones of the lower extremity and the hip. The australopithecines also used the limb bones of antelopes as implements. The fact that many baboon skulls have been found associated with australopithecine remains, in such condition as to suggest that they were bashed in from the top, has led Professor Raymond Dart of South Africa to conclude that this was the doing of australopithecines. While some authorities agree with this conclusion, others do not.

THE MOST PRIMITIVE KNOWN FORM OF MAN
The most primitive, though by no means the oldest, known form of man is a member of the subfamily Australopithecinae. This is *Zinjanthropus boisei,* discovered in July 1959 by Dr. L. S. B. Leakey and his wife in Olduvai Gorge, Tanganyika Territory, East Africa. "Zinj" is the classical Arabic name for East Africa and "anthropus," of course, means man. The spe-

cific name is in honor of an English benefactor of the expedition which led to the discovery.

Zinjanthropus boisei or Oldoway man is represented by an almost complete skull and part of a shinbone. The skull is that of a youth between sixteen and eighteen years of age, with a cranial capacity of 530 cc. The face is long and wide, there is a sagittal crest, almost no brow, large premolars and molars, but small canines and incisors. The skull was found on a living floor associated with stone tools of an early type known as Oldo-

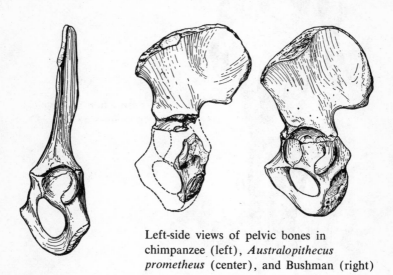

Left-side views of pelvic bones in chimpanzee (left), *Australopithecus prometheus* (center), and Bushman (right)

wan. These tools, made of quartzite and lava, are characterized by having only a few flakes removed in either one or two directions on both faces to make a simple chopping implement with a sharp but irregular cutting edge on either side of the edge of the stone. Similar tools made from waterworn pebbles have been found in association with three teeth at Sterkfontein in South Africa. First attributed to *Zinjanthropus*, Leakey now believes that these artifacts are the handiwork of another early man, to be described below (p. 47). Since there is no stone at the Olduvai I level at which these artifacts were found, the materials from which the tools were made must have been brought from elsewhere. The presence of 176 flakes indicates that the tools were manufactured on the spot.

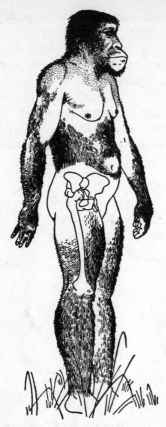

Reconstruction of a typical
australopithecine

Zinjanthropus, according to the potassium-argon dating (see p. 92) arrived at by Drs. G. H. Curtis and J. Evernden of the University of California, dates back 1.75 million years.

Most authorities are now agreed that *Zinjanthropus* is an australopithecine of the same type as *Paranthropus* from two sites in South Africa, Kromdraai and Swartkrans, and from another as far away as Java (*Meganthropus*).

"HOMO HABILIS"

In December 1960 Leakey reported further discoveries from the original *Zinjanthropus* site. These consisted of some cranial fragments together with a lower premolar and an upper molar tooth much more manlike than those of *Zinjanthropus,* and a shinbone and fibula.

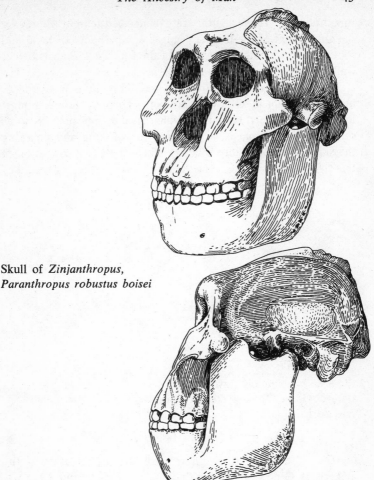

Skull of *Zinjanthropus,*
Paranthropus robustus boisei

Some 300 yards from the site of the discovery of *Zinjanthropus,* at another site and at a 2-foot lower level of Bed I, the Leakeys found, in the summer of 1960, the remains of at least two individuals, one juvenile, the other adult. The juvenile remains consist of portions of the parietal bones, part of an occipital bone and other skull fragments, a lower jaw, an upper molar, and parts of the skeleton of the hand. The adult bones are represented by a clavicle, some hand bones, and an almost complete skeleton of the left foot. The bones of the juvenile are those of a child of about 12 years. The skull bones are quite thin, and there is no sagittal crest or marked temporal lines for the attachment of the temporal muscles. The bones of hand and foot are very manlike, and those of the foot show that its owner walked and ran very much as modern man does.

Found in association with these skeletal remains were a large number of typical Oldowan artifacts. These included a most interestingly shaped bone tool, which Leakey interprets as some sort of "lissoir," that is, a tool used for working and polishing the skins of animals into usable leather. If the interpretation of the significance of this bone tool is correct, then, as Leakey

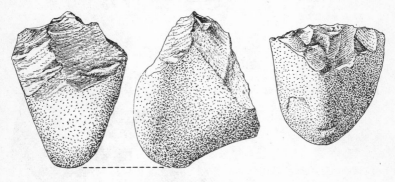

Oldowan pebble tools (lava), Bed I,
Olduvai Gorge, Tanganyika Territory,
East Africa

remarks, "It postulates a more evolved way of life for the makers of Oldowan culture than most of us would have expected." Great quantities of the remains of tortoises, catfish, and relatively easily caught aquatic birds were found at this level, suggesting that at this stage these early men had not yet progressed to the gathering and killing of the juveniles of larger animals. If this was so, how then account for the presence of a lissoir? A lissoir would be used for preparing the skins of large animals by smoothing out the roughnesses. Such special preparation of animal skins could only mean that they were used for domestic purposes, for clothes and as mats to sleep on, and even as covers for dwellings.

If these early men did not hunt big game, how then did they

secure the skins of such animals? Most probably they obtained them from the carcasses of animals that had either died a natural death or been killed by some other animal. It is also quite possible that they may have secured their animal food by scavenging. Vultures would have been the means of indicating to these early men where a carnivore had probably made a kill or an animal that had died in some other manner was to be found. This was one way of obtaining meat that has been practiced by many African peoples, among them the *strandlooping* Hottentots, down to the present day.

The fact that the bones of large animals have not been found upon the living floors of the early Oldowans suggests that if they did upon occasion scavenge, they ate the meat of the dead animal on the spot where it was found as, again, is frequently done by many African peoples to this day, especially when the animal is a large one like an elephant or a rhinoceros. If any of the meat had been carried back to their home territory it would have been encumbered with as little bone as possible, and probably cut into smallish pieces. This would explain the absence of the bones of large animals on the Oldowan living floors.

The cranial capacity of the 12-year old pre-zinj child has been calculated at about 680 cc, possibly more, and the age of the bones has been calculated by the potassium-argon method at 1.85 million years.

In October 1963 the Leakeys found fragments of a skull of late adolescent age and of the same physical type as the pre-zinj child. The pre-zinj child remains were found in Bed I, whereas the late adolescent remains were found in the lower part of Bed II. The remains, mostly of teeth, of some five other representatives of this type were also found in Beds I and II. Some of these finds occurred on the *Zinjanthropus* living floor. The shinbone and fibula may have belonged to this more advanced form.

The parts of the late adolescent remains consisted of most of the occipital bone, the two parietals, parts of the frontal and temporal bones, parts of the upper jaw, and a virtually complete mandible with all the teeth in place. In all physical features these remains closely resemble those of the pre-zinj child found in Bed I.

Because of dental size and shape, brain size, skull form, and specializations of the upper and lower extremities, and the association of manufactured stone artifacts, these remains have been described as representing a distinct advance on the australopithecines, and been given the name *Homo habilis* ("habilis" means able, handy, vigorous, mentally skillful), but most authorities believe that these remains are of australopithecines.

Fragments of a skull found in 1949 at Swartkrans, South Africa, probably of Middle Pleistocene age, and named *Telanthropus,* and some found in 1960 at Lake Chad, Koro Toro, north central Africa, of Late Lower or Early Middle Pleistocene age, may be "habiline," but until the necessary comparative studies have been carried out their exact status remains uncertain.

In January 1964 Leakey discovered an almost complete zinjanthropine mandible with all its teeth in place at a new site west of Lake Natron, northeast of Olduvai. The deposit is of the same age as the *Zinjanthropus* site.

The earliest known form of man thus far discovered is the australopithecine *Homo habilis,* though the morphologically most primitive is the australopithecine *Zinjanthropus.* The fact that a more developed form of man happens to have been discovered in an older stratum than a less developed form is almost certainly due to the accidents of archeologic serendipity. Stone tools of Oldowan type have been found at Sterkfontein and Swartkrans in association with australopithecines. Raymond Dart believes that most of the tools used by australopithecines were adapted from the skeletal remains of antelopes and similar animals and consisted of bone, teeth, and horn. Hence, he called it the *osteodontokeratic culture.*

Several remarkable tools, thought to have been made from animal bones by australopithecines, have been described by Dart. One such tool is an "apple-corer" found in association with *Australopithecus prometheus* at Makapansgat, South Africa. Another is a lissoir found in 1958 in the red-brown breccia at Sterkfontein, South Africa. The australopithecines may have learned to utilize such natural products as bone, teeth, and horn for the production of serviceable tools for digging, cutting, sawing, scraping, and smoothing.

The discovery of a rough circle of loosely piled stones on the living floor in the lower part of Bed I (lower and earlier than the zinjanthropine site), with hundreds of stone tools all around, in a region in which stones do not naturally occur, indicates that the habilines may have built shelters. Thus, in addition to possibly having been the first habitual makers of stone tools, they also may well have been the first builders of domestic dwellings.

At a third site at Olduvai in Bed I at a level some twenty feet above the other two floors, a remarkable assemblage of fossil bones of animals, many of them new to science, were found. All the large animal bones had been broken open and their marrow extracted. The skulls and jaws had been smashed. Most of them were the bones of immature animals.

Unlike their more apelike ancestors, both zinjanthropines and habilines were meat-eaters. Leakey has suggested that they may have obtained their meat by driving herds of animals into swamps and butchering on the spot those which were easily caught.

THE DEVELOPMENT OF HUNTING
AND ITS CONSEQUENCES

The finds at Olduvai enable us to follow, virtually step by step, the evolution of man's food habits and his methods of acquiring food. Three main stages may be recognized: (1) the extension of the habit of food-gathering, consisting almost entirely of plant foods, to the gathering of animal foods such as the slow-moving, easily caught animals, including tortoises, catfish, and aquatic birds, followed by the extension of this to (2) the collection of the young and larger animals, and, finally, (3) the hunting of the larger animals themselves.

The passage from a vegetarian to a meat diet marked a step in cultural evolution which probably had an important influence on the physical evolution of man. Meat requires much less chewing than fibrous plants. Large canine teeth are helpful in shredding the harder coverings of many edible plants, and bony crests are necessary to give attachment to the massive muscles that move the lower jaw of the plant-eating primate. For the meat-eater large canine teeth, large jaws, and bony crests are unnecessary. Hence, the changes that are destined to occur are toward the development of a more manlike head, with plenty of room for an expanding brain.

It is an interesting fact that meat-eaters have developed a far wider range of behavioral capacities than herbivores, presumably because they are forced to solve many more problems and meet the challenges of all sorts of situations with which herbivores are not confronted. For the herbivores in the forest the table, as it were, is laid, and all they need do is eat therefrom. To move from the gathering or hunting of small animals to the hunting of larger ones would have constituted a process of gradual development. In the course of hunting the advantage would be with those individuals who could stalk their prey and do so in the erect position. Effortless walking was probably achieved after the development of skilled running, and walking almost certainly served to put the selective pressure on those individuals who happened to possess the right genetic potentialities for the development of upright bipedal posture and locomotion. The upright posture frees the forelimbs for purposes other than those of locomotion, not only for the making of tools but also for their more efficient use in connection with the hunt.

Hunting small game puts a premium upon problem solving rather than upon biologically predetermined automatic reactions or instincts. In the savanna environment those individuals would have been favored who more often made the appropriate responses to the challenges of the environment than those who did not. Automatic *reactions,* instincts, would have been at a disadvantage while problem solving, which is another word for intelligence, would have been at an advantage. As such creatures fashioned their tools, made their snares, and dug the pits for their prey to fall into, intelligence would assume increasing value for survival. In this manner, in this new environment, by natural selection, early men would have continued to grow and develop in intelligence.

A creature that loses its instincts and has to rely increasingly upon intelligence for its survival must do two things: (1) it must develop an extended dependency period during which it must learn the basic essentials with which to function as a human being, and (2) it must develop a large and sufficiently complicated storage-retrieval repository, that is, a brain large enough to house the many billions of cells and their circuits necessary to serve such an intelligence.

As a consequence of the loss of any instincts it may once have possessed such a creature must learn everything he comes to do and know as a human being from other human beings. Hence, there develops a prolonged dependency period during which the fundamental part of this learning must occur. The warehouse for the necessary learning, storage, and retrieval of all this information must be large. In preparation for all this learning, therefore, toward the last month of gestation the brain of the human fetus grows at an accelerating rate, so that by the time it reaches between 350 and 400 cc in volume the baby must be born, otherwise its head would grow too large and it could not be born at all. And so at an average age of 266½ days from conception the child is born in an extremely immature condition. Indeed, when the human infant is born it has completed only half its gestation. That part which it has spent in development within its mother's womb is called *uterogestation.* The other half of its gestation the human infant must complete outside the womb in the continuing symbiotic relationship with its mother. This second half of its gestation is called *exterogestation.* Exterogestation lasts about as long as uterogestation, that is, about ten months, the age at which the infant generally begins to crawl about by himself.

The immaturity of the human newborn placed a high premium upon those women who were most capable of ministering to the needs of the dependent infant. And so those women would have been naturally selected who possessed this ability, or mother love, whereas those women who did not possess it would not have been as successful in leaving progeny behind. A high premium would also have been placed upon those men who exhibited cooperative qualities in the hunt and in social life generally, while the selfish, uncooperative individuals would not have fared so well.

It seems probable, then, that the pressures of environmental change first led to the development of those adaptive traits which transformed an apelike creature into a man. The most important of those traits are not so much physical as they are functional, behavioral, and those extra-corporealized traits that we know as *culture*—the man-made part of the environment, man's principal means of adaptation to the environment.

The functional traits that have evolved in interdependent relation in the human species, in addition to the adoption of the erect posture and the development of a large brain, have been the loss of what may have remained of instincts, the replacement of this reactive form of behavior with greatly maximized responsive problem-solving behavior, that is, intelligence, the development of a large and complex brain, birth in an immature and highly dependent state, a long dependent infancy during which the child lays the foundations of that knowledge necessary for him to become a functioning member of society, and the development of altruistic or cooperative behavior.

In other words, the loss of instincts, dependency and interdependency, altruism and cooperation, educability and intelligence have evolved in interrelation to produce the human species. It is with the uniquely high endowment of these potentialities that the human baby is born.

By becoming an omnivore man immeasurably increased his capacity for survival in all environments. He is among the few animals that are omnivorous. As an omnivore man can eat and digest virtually everything that is edible or that can be rendered so. This is one of the principal reasons why man has become the most widely distributed of all the creatures on earth.

With the advent of the australopithecines and their toolmaking activities the group of primates which eventually led to modern man move into a completely new zone of adaptation, the dimension of *culture*. With this transition there appears, for the first time, the essentially and characteristically human trait: the human mind. The human mind is characterized by an increasing capacity for the utilization of complex symbols, their application to the creation of novelty, and the solution of complex problems. This is accompanied by a growth in the ease and skill with which the symbols and what they are intended to represent are utilized. For the first time an animal, in a substantive manner, transcends its bodily limitations and the inarticulateness of nature, to become critical, analytic, and increasingly controlling of both.

RECONSTRUCTION OF THE COURSE
OF HUMAN EVOLUTION

There is some evidence that during the Pliocene epoch (which lasted some 10 million years) there was a gradual change in the climate of Africa. This consisted principally in a progressive withdrawal of rainfall from the south toward the equator. As a consequence the densely forested part of Africa below the equator gradually became deforested and transformed into open plains or savannas. The members of the stock which led to man, who formerly lived in forests, were in this manner gradually challenged to adapt themselves to the changing environment and the new problems which it presented.

In the forest plant food was abundant and hardly any effort was necessary to obtain it. On the open plains, however, it became increasingly sparse. In the new environment the vegetation was insufficient to sustain life. Hence, the immediate precursors of man would have been forced to supplement their diet by the gathering of small, infant, and slow-moving animals. From plant-eating they would have been forced to include also the eating of animals and thus to adopt an omnivorous diet.

A CHINESE MANLIKE APE: *Gigantopithecus blacki*
In China it has been an age-old custom to dig up so-called "dragon bones," which are in fact the fossil bones of many different kinds of animals, and to sell these, either whole or ground up, to the apothecary shops which in turn sell them as a potent medicine to willing buyers as cures for practically every disorder and disease. It may seem strange that a geologist should consider it part of his job to explore the contents of Chinese apothecary shops in search of fossil material, but when we learn that it was by so doing that the Dutch geologist G. H. R. von Koenigswald discovered the tooth of one of the most primitive manlike forms known to scientists, it will not seem so strange after all. Actually Dr. von Koenigswald discovered three manlike molar teeth. These consisted of right and left lower third year molars belonging to different individuals, and

an upper molar probably belonging to still a third individual. The striking thing about these teeth is their size. They are enormous. The volume of the crown of the lower third molar is about six times larger than that of the equivalent tooth in modern man, and it is almost twice as large as the corresponding tooth of the gorilla.

Authorities like the late Professor Weidenreich considered *Gigantopithecus* to be a giant man. Both the gigantism and the ascription of the teeth to a man had been questioned. With the discovery in 1957 of a *Gigantopithecus* mandible, with most of the teeth *in situ,* in a Middle Pleistocene deposit in a high cliff cave in Kwangsi Province, south China, it is now clear that *Gigantopithecus* was an ape, an advanced ape, but certainly not an ape-man or in the line of man's descent.

In fact the creatures to which these remains belonged were tremendously robust apes, but there is no reason to believe that they were giants or, as some authorities had suggested, that they were related to the australopithecines of Africa. Von Koenigswald, however, believes that they have closer affinities to man than any other known ape, living or extinct.

A JAVANESE MANLIKE FORM:
Meganthropus palaeojavanicus
One of the oldest manlike forms known (*Meganthropus palaeojavanicus*) was discovered in 1941 by Dr. von Koenigswald in the form of the fragments of two lower jaws in the Lower Pleistocene beds of the Sangiran district in central Java. In 1952 a more complete, though badly crushed, jaw of *Meganthropus* was found at Sangiran by Dr. Pieter Marks. These *Meganthropus* jaws are extraordinarily massive, achieving the proportions of the jaw of an adult male gorilla. Nevertheless, the form of the jaw is distinctly human. Dr. von Koenigswald regards this genus as ancestral to *Homo erectus.*

Homo erectus erectus
During the years 1890 through 1897 a young Dutch physician named Eugene Dubois, who had gone to Java in search of "the missing link," discovered at Trinil in central Java the top of a skull, a thigh bone, the fragment of a lower jaw, and three teeth. These were all strikingly manlike, though the skullcap still

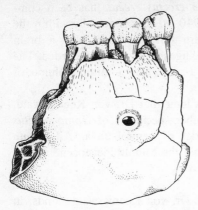

Fragment of the right side of the lower jaw of the so-called *Meganthropus palaeojavanicus,* but clearly a member of the australopithecines substage *Paranthropus,* from the Lower Pleistocene of Sangiran (central Java) (Courtesy of Professor G. H. R. von Koenigswald)

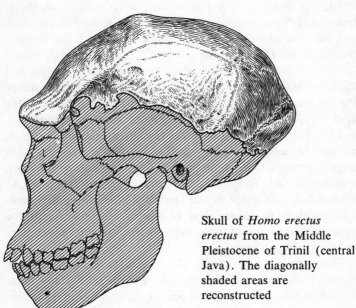

Skull of *Homo erectus erectus* from the Middle Pleistocene of Trinil (central Java). The diagonally shaded areas are reconstructed

looked pretty primitive. The thigh bone was almost like that of a modern man, suggesting that the creature to which it belonged had walked erect. This combination of apelike top of the head and manlike thigh bone suggested the name "the ape-man who walked erect," hence, in its Latin form, *Pithecanthropus erectus,* but today better known as *Homo erectus erectus.*

The brain volume of *Homo erectus erectus* has been computed to be between 775 and 940 cc. This is actually within the range of modern man. Modern European men with a brain volume of 850 cc with good intelligence have been recorded. The great French writer Anatole France had a brain volume of slightly over 1000 cc.

It was not until many years later that Dr. von Koenigswald, in 1937, discovered several other specimens of *Homo erectus erectus* in the Sangiran district of central Java. All *Homo erectus erectus* remains are from the Middle Pleistocene epoch.

Homo erectus robustus

One of the most important of Dr. von Koenigswald's finds, in 1939, was that of a very robust form of *Homo erectus,* therefore called *Homo erectus robustus.* This consisted of the back and the base of the skull and the upper jaw with the teeth in their sockets. The teeth are essentially human in form, except that the canines project beyond the level of the other teeth, and there is a bony space between the canine and the lateral incisor for the reception of the tip of the lower canine, just as in the apes. In *Homo erectus erectus* this space has disappeared.

The evidence, therefore, points to this evolutionary order of changes in the jaw region, namely, that first the canine tooth underwent reduction, as we can see in *Homo erectus robustus,* and that this was followed after a time by disappearance of the premaxillary diastema, as the space between the canine tooth and the lateral incisor is called. In this way the projection of the upper jaw was reduced to the more or less straight form characteristic of modern man.

Early in December 1960 Dr. L. S. B. Leakey discovered, some 15 to 20 feet below the top of Bed II at Olduvai, the calvarium (skull without facial bones or jaws) of a pithecanthropine of *Homo erectus erectus* type. The importance of this discovery is that, first, it extends the range of the pithecanthropines, and second, the presence of this pithecanthropine at Olduvai suggests that this type of man may very well represent a more evolved form originating from the habilines. About 100 yards away from the site of the discovery, and at the same level, were found numerous hand axes of Abbevillian industry, as well as the bones of large animals which appear to have been

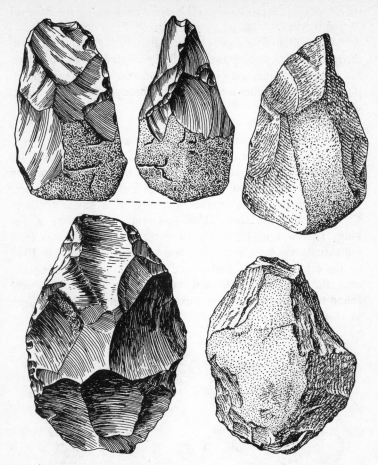

Abbevillian (Chellean) stone tools

broken open for their marrow. Potassium-argon tests yield the date of 490,000 years for these remains.

On the Olduvai Abbevillian living floor were found many large bola stones. The bola, a hunting weapon of great ingenuity, consists of three stones held in leather bindings by three thongs which are knotted together at the top, leaving free the thongs and the stones fastened to them. The bola is whirled around the head of the hunter, who then aims it at his prey's legs, around which the weighted thongs fasten and thus bring the entangled creature down. It is a device used by the gauchos of South America and by some Eskimos to this day. The size of the Abbevillian bola stones suggest that the Olduvai pithecanthropine was of very powerful physique. Red ocher found on his living floor indicates his interest in color for some special purpose, for the ocher must have been brought from a considerable distance.

Sinanthropus pekinensis: Homo erectus pekinensis
Sinanthropus pekinensis means Chinese man from Peking. This name was given to what he considered to be a new form of man by Professor Davidson Black, then of Peking Union Medical College, on the basis of the discovery of a single tooth at Chouk'outien, just about thirty-seven miles southwest of Peking. This was in 1927. By 1939 the remains of over forty individuals had been recovered, and new digging begun in 1943 on the original Middle Pleistocene site has succeeded in uncovering the remains of additional skeletal parts. The remains were found to represent a Chinese form of *Homo erectus,* now known as *Homo erectus pekinensis.*

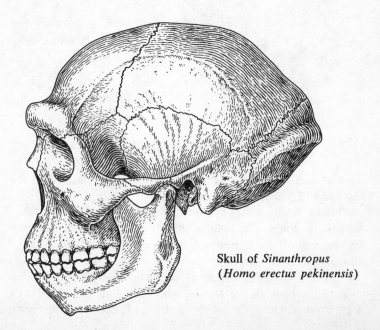

Skull of *Sinanthropus*
(*Homo erectus pekinensis*)

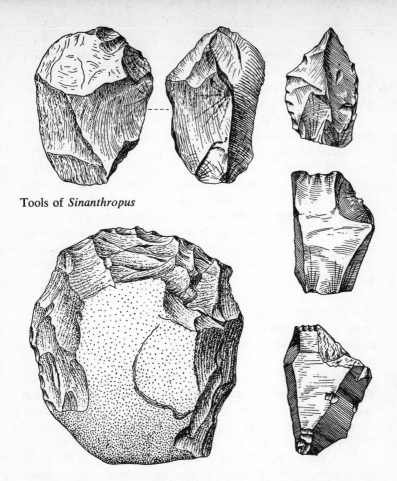

Tools of *Sinanthropus*

Unfortunately the original skeletal remains of Peking man were lost while they were being transported to an unachieved safety during the Japanese invasion of China. Casts, however, of most of the original material are available as well as good photographs and drawings.

The average brain volume of Peking man is 1075 cc. Brain size is about 20 per cent greater than that of Java man. The forehead region is slightly more developed in Peking man than in Java man. The teeth are human in form, and there is no space between the canine tooth and the lateral incisor in the upper jaw, but, as in the pithecanthropines, a developed chin is lacking.

Many tools of the chopper and cutting types were found associated with Peking man.

It has been asserted that Peking man was a cannibal because the bases of all the skulls were found broken open, and many of the long bones had been split longitudinally by some human means. The suggestion is that *Sinanthropus* extracted the brain and ate it, and also sucked on the long bones to extract the marrow. This is possible but unproven, and in any event would not necessarily imply that *Homo erectus pekinensis* habitually practiced cannibalism. Under conditions of extreme starvation most human beings are capable of cannibalism. Except in aberrant cases it is highly unlikely that man has ever resorted to eating his own kind except *in extremis* or for ritual purposes.

The resemblances between Java and Peking man are striking, and all in all it seems justifiable to conclude that the latter represents a slightly more advanced geographic variety of the former, and is to be included among the pithecanthropines.

Homo erectus mauritanicus

In June 1954 Professor C. Arambourg of the National Museum of Natural History at Paris discovered two human lower jaws in a pit at Ternifine in Algeria. These jaws came from a Lower Middle Pleistocene (Kamasian industry) horizon, to which Professor Arambourg attributes an age of about half a million years. The jaws, which are very robustly built, are without developed chins and strongly resemble those of the pithecanthropines, but are perhaps sufficiently different from them to warrant identifying them as a mauritanian variety of the type, *Homo erectus mauritanicus* rather than *Atlanthropus mauritanicus,* the name given to this type by Professor Arambourg.

The teeth in *mauritanicus* are unmistakably human and most closely resemble those of the pithecanthropines. Associated with these jaws were found the remains of many extinct animals and numerous roughly worked stone tools of quartzite, limestone, and flint, representatives of the most ancient type of workmanship (Abbevillian-Acheulian).

A portion of a human lower jaw also found in 1954, in a Middle Pleistocene deposit at Sidi Abderrahman near Casablanca in Morocco, associated with stone tools of a Middle Acheulian industry, appears to belong to the same type as *mauritanicus*. Thus, for the first time we have definite evidence of the presence of pithecanthropine types in Africa. Earlier, in

1934, the fragments of three skulls discovered in an Upper Paleolithic deposit northwest of Lake Eyassi, in East Africa, were by some authorities thought to be of pithecanthropine type. The type represented by these skull fragments was named *Africanthropus*. The new discoveries in North Africa strengthen somewhat the claim of *Africanthropus* to pithecanthropine status.

SOLO MAN

Near Ngandong in central Java in 1931, in the region of the Solo River, there were found eleven fossil skulls. Faces and teeth were missing, but the skulls, which were extraordinarily thick, showed distinct resemblances to *Homo erectus* on the one hand and to the later forms of man known as Neanderthal on the other. The average brain volume was 1100 cc.

With the remains of Solo man were found examples of his handiwork in the form of several beautifully worked bone implements, an ax made of deer antlers, a barbed spearhead, and a number of somewhat crudely fashioned stones. Solo man

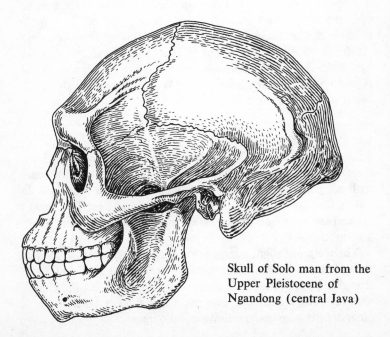

Skull of Solo man from the Upper Pleistocene of Ngandong (central Java)

appears, then, to have been an advanced form culturally, belonging somewhere in the Upper Paleolithic or Upper Old Stone Age (see Table 4).

Direct association of such implements with *Homo erectus* has never been demonstrated, but throughout the Sino-Malayan region there are found a variety of chopping tools which may have been the handiwork of the pithecanthropines.

WADJAK MAN

A very interesting discovery was made by van Rietschoten at Wadjak, some sixty miles southeast of Trinil in central Java, of two human skulls during the years 1889 and 1890. The announcement of the discovery of these skulls was not made until 1922, by Dubois. Dubois claimed a Pleistocene age for these two skulls, the brain volume of one being 1550 cc and of the other 1650 cc. The skulls are probably of late Pleistocene age, and are clearly of *sapiens* type.

What is so interesting about these skulls is that they bear a remarkable resemblance to the skull of the typical Australian aboriginal of today, except that they have a larger brain volume than the latter, with an average of about 1300 cc. It is quite possible that some members of the Wadjak population reached Australia in Pleistocene times. It has been suggested that we now have an almost continuous evolutionary line leading from *Homo erectus* through Solo man to Wadjak man, and thence to the Australian aboriginal.

HEIDELBERG MAN

In a quarry at Mauer, some six miles southeast of Heidelberg, Germany, the massive lower jaw of a primitive type of man, with all the teeth in place, was uncovered by a workman in 1907. This jaw is of Lower Middle Pleistocene age. This makes Heidelberg man one of the oldest authenticated human fossils known to us.

The teeth are slightly larger than those of the average man of today but well within modern man's normal range of variation. The side of the jaw (the ramus) is very broad, and the chin is undeveloped. Heidelberg man may represent a forerunner of Neanderthal man and a relative of Solo man.

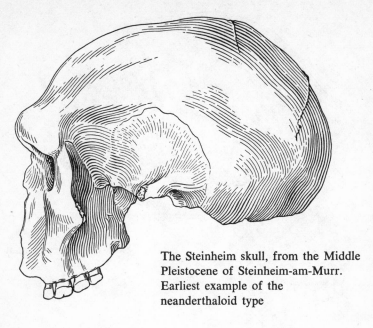

The Steinheim skull, from the Middle
Pleistocene of Steinheim-am-Murr.
Earliest example of the
neanderthaloid type

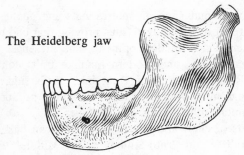

The Heidelberg jaw

RHODESIAN MAN

A complete skull minus a lower jaw together with some post-
cranial bones and the parietal and maxilla of another individual,
all of Upper Pleistocene age, of a primitive type of man was
found in a cave at Broken Hill, in Northern Rhodesia, now
known as Zambia, in 1921. Combining neanderthaloid traits
with modern manlike traits, Rhodesian man had a brain volume
of 1280 cc, massive brow ridges, a projecting, large upper jaw,
and an unusually broad palate. The teeth closely resemble those

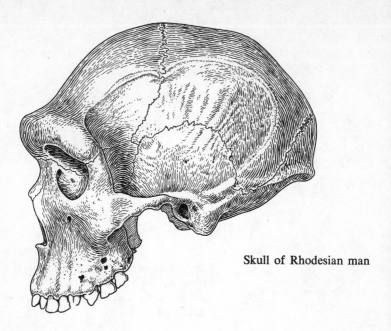

Skull of Rhodesian man

of contemporary man, and an interesting thing about them is that they were all badly decayed, thus proving that bad teeth are not altogether a modern development. The presence of Solo-like, Neanderthal-like, and modern manlike traits combine to render the Rhodesian skull of great interest. It suggests that Rhodesian man may actually be an evolutionary product of the admixture, among other things, of such types.

Implements of chert and quartz of the African flake culture known as Stillbay and Proto-Stillbay of the Middle Stone Age were found adjacent to the remains.

In 1953 another skull of Rhodesian man, together with associated artifacts (objects made by man), was found about fifteen miles from Saldanha Bay, some sixty miles north of Cape Town, in South Africa, and some fifteen hundred miles from Broken Hill where the first skull was found. This shows that Rhodesian man roamed pretty widely over Africa.

Only the base and face and lower jaw are missing from the skull, but the skullcap with its sides, back, and great brow ridges shows quite conclusively that we are here dealing with

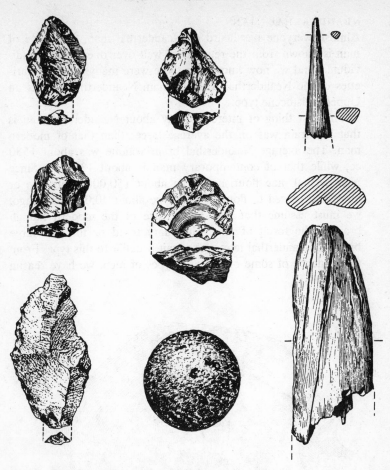

Tools found in association with Rhodesian
man (Courtesy of the British Museum
[Natural History])

the same Solo-Neanderthal-modernlike medley of traits that
characterized Rhodesian man I.

In Rhodesian man, then, we have perhaps a link between
Solo man of Java, Neanderthal man of many parts of the world,
and the modern, or neanthropic, type of man, as he is sometimes called.

NEANDERTHAL MAN

Almost everyone has heard of Neanderthal man. This type of man is known from the remains of well over one hundred individuals, and we now know that there were many different varieties of the Neanderthal form of man. Neanderthal man is an Upper Pleistocene type.

The first thing of interest to say about Neanderthal man is that his brain was on the average larger than that of modern man. The average Neanderthal brain volume was about 1550 cc, while that of contemporary man is about 1350 cc. Since Neanderthal man flourished from about 150,000 years ago or more and ceased to flourish as a type about 40,000 years ago, we must assume that the smaller size of the modern human brain is the result of an evolutionary trend or that the large brain of Neanderthal man was a trait peculiar to this type. From the evidence of some other early types of men, we have reason

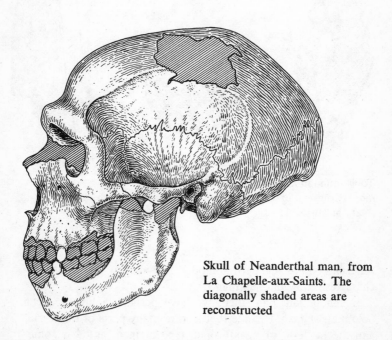

Skull of Neanderthal man, from La Chapelle-aux-Saints. The diagonally shaded areas are reconstructed

to believe that the human brain has actually decreased in gross size and stabilized itself about 50,000 years ago at its present size. In spite of the prognostications of writers in Sunday magazine supplements, it is highly unlikely that the human brain will evolve by growing larger. It doesn't have to. It can increase in complexity without increasing in size. It can enlarge its surface area without growing in volume by the increase and deepening of its convolutions.

Neanderthal man tended to have a somewhat sloping forehead, with well-developed brow ridges, a heavy chinless jaw, and a rather projecting back of the head (occiput). Owing to the want of a little knowledge of elementary anatomy, some "authorities" who engaged in the "reconstruction" of Neanderthal man represented him with a bull neck, grotesque features, and walking with a stoop, during which, it was alleged, his knees knocked together! It has also often been asserted that Neanderthal man must have been of low intelligence because he had a low forehead. All these statements are quite indefensible. Neanderthal man walked as erect as any modern man, he did not have a bull neck, and he was not knock-kneed. And it has long ago been proven by many independent scientific investigators that neither the form of the brow nor the shape of the head have anything whatever to do with intelligence. As a matter of fact, Neanderthal man's brow was quite well developed. What makes it appear low is the presence of massive eyebrow ridges (supraorbital tori). There is, in fact, every reason to believe that Neanderthal man was every bit as intelligent as contemporary man. He made the beautiful tools which are described as belonging to the Mousterian culture (after Le Moustier in southern France where they were first found). He made flint balls, perforators, discs, scrapers, and stone knives, and developed the use of mineral pigments such as red ocher for ceremonial and possibly other purposes. Neanderthal man also introduced ceremonial interment of the dead, thus suggesting that he possessed a highly developed religious system.

For a long time it was believed that Neanderthal man was exterminated by men of our own type. There was absolutely no evidence for this belief except the kind of theorizing that characterized nineteenth-century thinkers of "The Survival of the

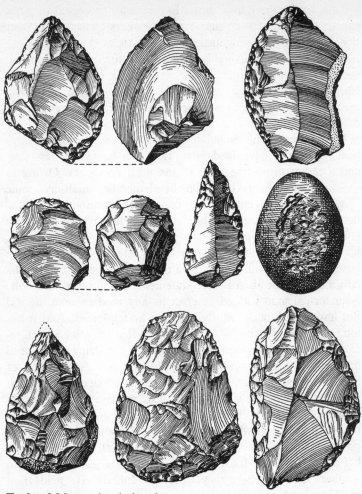

Tools of Mousterian industries

Fittest" school who believed that warfare was as old as man and was the means by which one race conquered and exterminated another. Since Neanderthal remains are known from almost every part of the earth where human fossils have been found, it is inconceivable that Neanderthal man should everywhere have

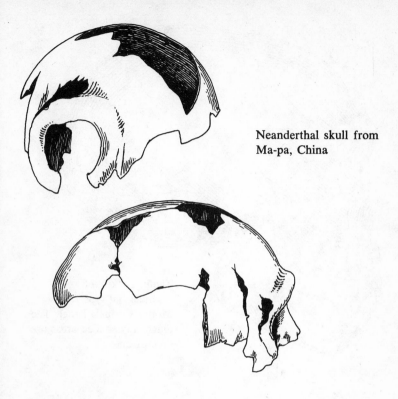

Neanderthal skull from
Ma-pa, China

been exterminated. It is doubtful whether he was exterminated anywhere. The truth seems to be that he mixed with whatever populations he encountered and in the course of time was absorbed by such populations. Certainly this was the case in the Middle East and in Europe, where many persons may still be seen who bear the traces of their remote Neanderthal ancestry. These traces may be observed in the rather heavy brow ridges, deep-sunk eye sockets, receding foreheads, and weakly-developed chin regions.

EARLY MIXTURES OF DIFFERENT TYPES OF MAN
While the populations of early man were very small, there is good reason to believe that whenever such populations met they did exactly what modern populations do, they interbred. Actual

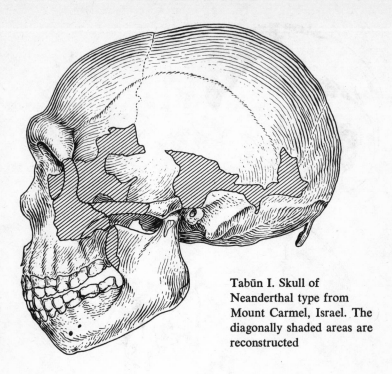

Tabūn I. Skull of
Neanderthal type from
Mount Carmel, Israel. The
diagonally shaded areas are
reconstructed

evidence of such intermixture was until recently a matter of
speculation, but during 1931 and· 1932 the evidence became
rather more strong. During this period an assemblage of fossil
neanderthaloids was discovered in caves on the slopes of Mount
Carmel in Palestine.

Here were found two types, a clearly Neanderthal type, in the
caves of Tabūn, and another type which closely approached
modern man, from the caves of Skhūl, a few yards away. Between the two types there was every variety of intergradation.
The evidence indicates that the remains from the two caves were
of broadly contemporaneous age. The radiocarbon age is 45,000
years B.P. (Before the Present). It is probable that there had

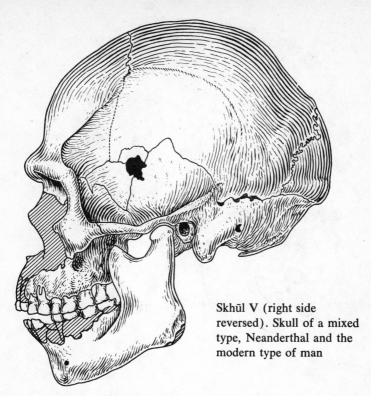

Skhūl V (right side reversed). Skull of a mixed type, Neanderthal and the modern type of man

been intermixture between a modernlike form of man and Neanderthal man and that the Mount Carmel population was the product of that intermixture.

There is every reason to believe that similar intermixture (hybridization) occurred between populations throughout the long prehistory of man.

CRO-MAGNON MAN

The Cro-Magnons are the Apollos of the prehistoric world. Cro-Magnon man was discovered originally in 1868 in the little village of Les Eyzies in southern central France, in a rock shelter called Cro-Magnon. The remains of thirteen other

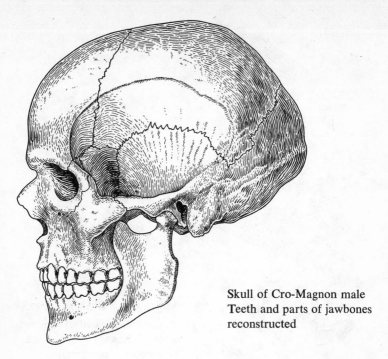

Skull of Cro-Magnon male
Teeth and parts of jawbones
reconstructed

individuals were uncovered between 1872 and 1902 in the caves of the Red Rocks of the Côte d'Azur, some forty minutes' walk from Mentone on the Italian Riviera. A headless, incomplete skeleton found in Paviland Cave in southwestern Wales in 1823 almost certainly belongs to the Cro-Magnon variety of man.

The Cro-Magnons were about 5 feet 11 inches in height, with a brain volume in the larger representatives of 1660 cc, a straight face, a well-developed, projecting nose, a high forehead, and a strong jaw. They made the most beautiful bone implements associated with an industry known as Aurignacian (from Aurignac in France where they were first found).

Cro-Magnon man is a modern man in every sense of the word, but where he came from or how he came about we have not the slightest idea.

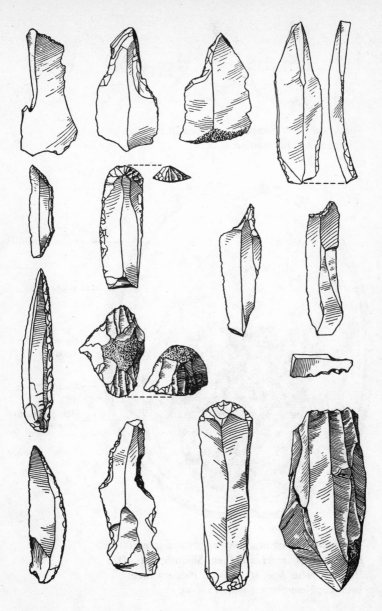

Upper Paleolithic flint tools

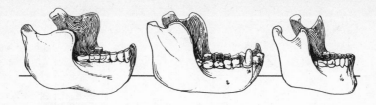

The lower jaws compared of Heidelberg man,
chimpanzee, and modern man

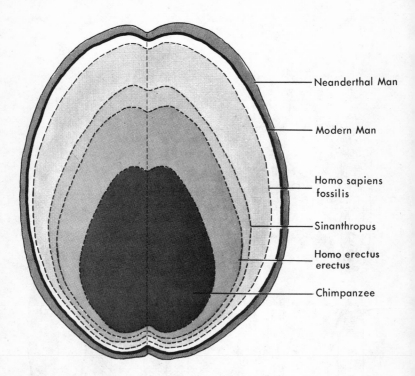

A comparison of brain sizes: chimpanzee 400 cc,
Homo erectus erectus 860 cc, Sinanthropus 1075 cc,
Homo sapiens fossilis 1300 cc, modern man
1400 cc, Neanderthal man 1550 cc

At right, a comparison of casts of the brains, the mid-sagittal
sections of the skulls, and the inner aspects of the lower
jaws of anthropoids and man

Brain-Cast
Gibbon

Lower Jaw
Chimpanzee

Brain-Cast
Chimpanzee

Skull-Section
Chimpanzee

Lower Jaw
Orangutan

Brain-Cast
Gorilla

Skull-Section
Homo erectus erectus
Face Restored

Lower Jaw
Primitive
Neanderthaloid
Ehringsdorf

Brain-Cast
Pithecanthropus

Lower Jaw
Heidelberg Man

Skull-Section
Neanderthal Man
La Chapelle-Aux-Saints

Brain-Cast
Neanderthal Man
La Chapelle-Aux-Saints

Lower Jaw
Neanderthal Man
La Chapelle-Aux-Saints

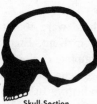
Skull-Section
Cro-Magnon Man

Lower Jaw
Cro-Magnon Man

Brain-Cast
Modern Man

Lower Jaw
Modern White Man

✹ 5 ✹

THE RISE OF MODERN
MAN

At one time, things used to be quite simple when anthropologists discussed the origin and evolution of man. Modern man was the last and highest type of man to be evolved. The most primitive type of man was held to be *Homo erectus* (*Pithecanthropus*) *erectus* or possibly *Homo erectus* (*Pithecanthropus*) *robustus*. All the types of men in between the pithecanthropines and modern man were held to be simply intermediate stages in the evolution of man from the most primitive to the most advanced type. However, certain finds of fossil man made in recent years render this earlier and apparently perfectly logical interpretation of the evolution of man seriously open to question.

SWANSCOMBE MAN

In 1935 and 1936 there were found in the Barnfield gravel quarry at Swanscombe in Kent, England, two skull bones, the left parietal and the complete occipital bone, of a single individual in a Middle Pleistocene deposit associated with tools of typically Acheulian industry. In 1955 the right parietal bone was discovered. The estimated cranial capacity is 1325 cc. The bones are somewhat thicker than those of modern man, the parietals somewhat flatter, and the occipital broader. Hence, Swanscombe man is considered to represent a form intermediate between paleoanthropic and neanthropic man. Of the same general type is Fontéchevade man.

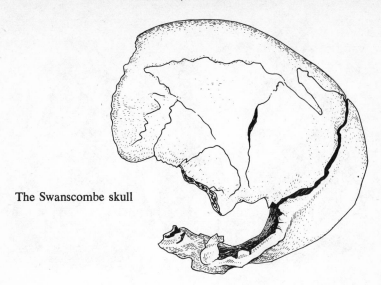

The Swanscombe skull

FONTÉCHEVADE MAN
In 1947 in a cave near the village of Montbrun in the Department of Charente, France, in a deposit of the Third Interglacial period, the portions of two skulls were discovered, associated with artifacts of Tayacian industry. The Tayacian industry always precedes the Mousterian industry which is associated with Neanderthal man. We can, therefore, be fairly certain that Fontéchevade man preceded Neanderthal man.

Fontéchevade I is represented by a beautifully preserved fragment of the region above the nose and the upper margin of the left orbit. This is a critically important part of the skull because it tells us at a glance what the forehead and facial region were like in the complete skull. In this type, almost all authorities agree, the face was exactly as in modern man. Fontéchevade I was possibly either of late adolescent age or a young adult female.

Fontéchevade II is represented by an almost complete skullcap. Facial bones and the base of the skull are missing. But again, except for thickness of bones, this is in every way a modernlike skull. The cranial capacity was about 1470 cc.

In Fontéchevade man we see yet another type with which the

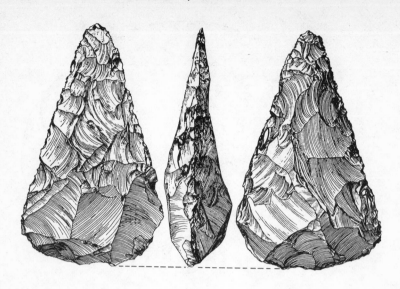

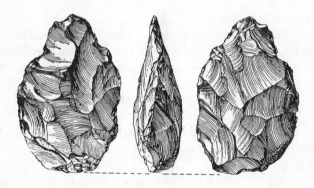

Acheulian hand axes associated with the Swanscombe skull

forerunners of Neanderthal man, and later Neanderthal man himself, may have intermixed.

KANAM MAN
In 1932 at West Kanam, on the southern shores of the Kavirondo Gulf of Victoria Nyanza, in Kenya, East Africa, the front portion of a human lower jaw was discovered. The

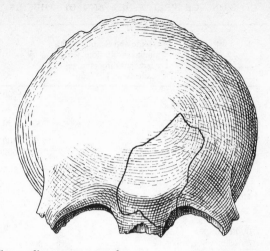

Fontéchevade I. The outline represents the
significant portion of the recovered skull

importance of this mandible lies in the fact that it is in every
way completely modern in appearance, yet it is claimed to have
been recovered from a Lower Pleistocene deposit (now unfortunately washed away), which would make the Kanam mandible very old indeed. However, there remains a doubt as to its
contemporaneity with the deposit in which it was found. Kanam
man suffered from a cancer of the chin region, which overgrew
that region in the form of a tumor, and therefore leaves in doubt
the question as to whether Kanam man had a well-developed
chin before it was obliterated by the tumor, though the form of
what little there is of the jaw strongly suggests that Kanam man
did possess a developed chin.

KANJERA MAN

At Kanjera in Kenya, East Africa, the skeletal remains of three
individuals of modern Negroid type were found in a Middle
Pleistocene deposit associated with implements of Acheulian
industry, that is to say of the same age and industry as
Swanscombe man in England. Here, again, the contemporaneity
of the remains with the deposit in which they were found has
been questioned.

TABLE 4. CHRONOLOGICAL-CULTURAL TABLE OF THE DIVISIONS OF PREHISTORY AND OF THE HISTORIC PERIOD

Stage	Alpine and Scandinavian glacial oscillations with corresponding changes of sea level and climate		Northwestern Continental Europe
POWER TOOLS		A.D. 1900	Rise of Age of power tools
NEW		A.D. 1850	Steel Age develops
STEEL		A.D. 1700	Steel (carbonized iron)
OLD		A.D. 1000	
	Present conditions of Mya Period in Baltic area	A.D. 500	Viking Age
Late		50 B.C.	Roman Period of Iron Age
Middle IRON		500 B.C.	Iron Age introduced
Early		1,000 B.C.	Northern Bronze Age
		1,500 B.C.	
BRONZE		2,000 B.C.	
Late	Final land rise in Baltic area or Late Tapes Period		Northern Neolithic
		2,500 B.C.	
Middle NEOLITHIC		3,000 B.C.	
Early		3,500 B.C.	Ertebølle Mesolithic
		4,000 B.C.	
		5,000 B.C.	
MESOLITHIC	Sea rising, Ragunda retreat, with Littorina Sea (Early Tapes period) preceded by late Ancyclus Lake	6,000 B.C.	Maglemose Mesolithic
		7,000 B.C.	
	Ragunda pause with Ancyclus Lake	10,000 B.C.	Tanged Point Cultures (Lyngby etc.)

Principal Culture Stages of Europe, Egypt, and the Near East

West Central Europe	Near East	Egypt	Human Types
La Tene Iron Age			
Hallstatt Iron Age			
Late Bronze Age (Urnfield)			
Middle Bronze Age (Tumulus)			
Early Bronze Age (Aunjetitz)			Persisting varieties of *Homo sapiens*
Late Neolithic (Danubian III, Western and corded-battle ax admixture)			
Danubian Neolithic II and Western Neolithic cultures			
1st movement of Danubian Neolithic peoples into Central Europe	Early Dynastic, Proliterale	Dynastic Egypt	
	Warka, Ubaid, Halaf	Nakada II (Gerzean)	
Tardenoisian, Azilian,	Hassuna, Matarrah	Nakada I (Amratian)	
Asturian and other Mesolithic cultures	Basal Mersin	Badarian	
	Early Neolithic, Jericho and Jarmo	Tasian	
	Natufian		Chancelade

TABLE 4 (*continued*)

Stage		Alpine and Scandinavian glacial oscillations with corresponding changes of sea level and climate		Northwestern Continental Europe
Upper		Fini-glacial pause with Baltic ice-lake	13,500 B.C.	
		Gothi-glacial retreat with Baltic ice-lake	15,000 B.C.	
		Gothi-glacial pause with Baltic ice-lake	20,000 B.C.	
Middle		Würm or Achen and Dani-glacial retreats with Frankfort and Pomeranian pause. Flandrian terrace	25,000 B.C.	
			40,000 B.C.	
		Würm and Brandenburg or Dani-glacial advances, 4th Glacial	50,000 B.C.	
			75,000 B.C.	
	PALEOLITHIC	Riss retreat with Monastrian terrace, 3rd Interglacial. Hot summer		
			150,000 B.C.	
		Riss and Polonian advances, 3rd Glacial	250,000 B.C.	
Lower		Mindel retreat with Tyrrhenian terrace, 2nd Interglacial		
		Mindel advance, 2nd Glacial	400,000 B.C.	
			500,000 B.C.	
		Günz retreat with Milazzian terrace, 1st Interglacial		
			600,000 B.C.	
		Günz advance, 1st Glacial. Pleistocene	1,000,000 B.C.	
		Villafranchian	2,000,000 B.C.	

Principal Culture Stages of Europe, Egypt, and the Near East

West Central Europe	Near East	Egypt	Human Types
Magdalenian	Atlitian	————	Předmost, Baker's Hole, Rhodesian, Cro-Magnon, Châtelperron, Grimaldi, Africanthropus, Solo
Solutrean	Aurignacian		
Aurignacian and Perigordian		Sebilian of Nile silts	
Mousterian	Levalloiso-Mousterian		Boskop
Acheulian, Levalloisian, Tayacian and Micoquian	Acheulian	"Egyptian Levalloisian" industry of 10 m. Nile terrace	Florisbad, Skhūl, Gibraltar II, Wadjak, Rhodesian
			Tabūn, Neanderthal, Ehringsdorf, Fontéchevade, Montmaurin
Derived implements			Rabut, Sidi Abderrahman
Abbevillian, Acheulian, Clactonian and Levalloisian	Acheulian	Acheulian industry of 50 m. Nile terrace	Steinheim, Swanscombe, Heidelberg, Atlanthropus Sinanthropus, Homo erectus erectus
Derived implements			Homo erectus robustus
Proto-Abbevillian industry from below Cromer forest beds	Abbevillian	Abbevillian and Early Abbevillian of 100 m. Nile terrace	Homo erectus erectus (Modjokerto), Meganthropus
			Australopithecus
			Australopithecus, Homo habilis

Comparison of skulls:
[A] male gorilla
[B] *Zinjanthropus*
[C] *Sinanthropus*
[D] Neanderthal
[E] modern

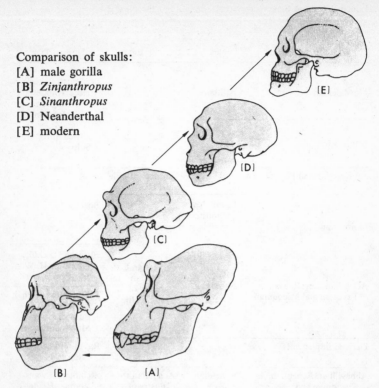

THE EARLY APPEARANCE OF NEANTHROPIC MAN

If Swanscombe, Fontéchevade, Kanam, and Kanjera man are representatives of the modernlike type of man, the conclusion is unavoidable that a type of man resembling modern man is much more ancient than was previously supposed. It is quite possible that Neanderthal man, for example, represents the product of admixture between such types as Solo with neanthropic man, rather than being a descendant of *Homo erectus* or Solo man in the direct line.

The older notion of straight-line (ortholinear) evolution from ape to increasingly advanced types of man is essentially sound, though in its oversimplified form of more advanced types appearing in a continuous straight line from less advanced types, it requires substantial modification.

Evolution is best likened not to a straight line but rather to a reticulum or network, in which all sorts of crisscrossing lines

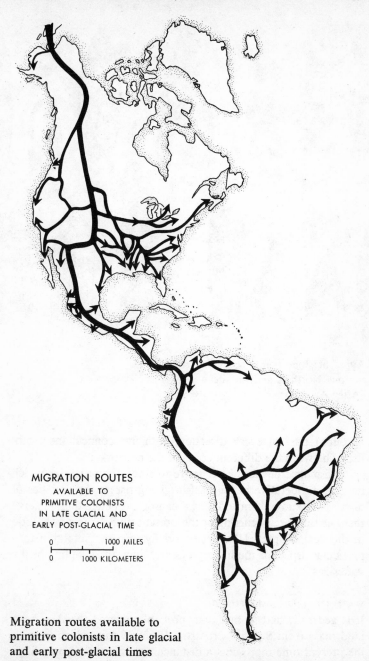

MIGRATION ROUTES
AVAILABLE TO
PRIMITIVE COLONISTS
IN LATE GLACIAL AND
EARLY POST-GLACIAL TIME

0   1000 MILES
0   1000 KILOMETERS

Migration routes available to
primitive colonists in late glacial
and early post-glacial times

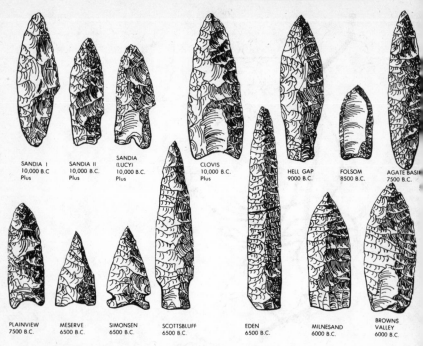

SANDIA I
10,000 B.C
Plus

SANDIA II
10,000 B.C.
Plus

SANDIA
(LUCY)
10,000 B.C.
Plus

CLOVIS
10,000 B.C.
Plus

HELL GAP
9000 B.C.

FOLSOM
8500 B.C.

AGATE BASIN
7500 B.C.

PLAINVIEW
7500 B.C.

MESERVE
6500 B.C.

SIMONSEN
6500 B.C.

SCOTTSBLUFF
6500 B.C.

EDEN
6500 B.C.

MILNESAND
6000 B.C.

BROWNS
VALLEY
6000 B.C.

Types of stone points recovered from
various North American sites
(After Agogino)

(groups) pass in every direction, with interconnections established between the different cords of the network.

As a matter of fact, if we desire to study the processes which have been operative in the past in giving rise to the varieties of men as we know them today, we can do no better than observe them as they are occurring at the present time. This we shall do in the next chapter. Here let us briefly turn our attention to a particular aspect of the history of early man, the peopling of the Americas.

MAN IN THE AMERICAS

It is generally agreed that man probably entered the American land mass from Siberia across the Bering Strait, which even at the present time represents a distance across water of only fifty-

TABLE 5. SOME NORTH AMERICAN SITES ASSOCIATED
WITH PREHISTORIC MAN

Site	Locality	Years Ago
Borax Lake	Borax Lake, California	4,000 to 7,000
Pinto Basin	Riverside County, California	7,000 to 9,000
Lake Mohave	Lake Mohave, southeastern California	7,000 to 9,000
Gypsum Cave	Frenchman Mountains, east of Las Vegas, Nevada	7,500 to 9,500
Scottsbluff Quarry	Scottsbluff, western Nebraska	8,000 to 10,000
Russell Cave	Jackson County, Alabama	8,160 ± 300
Simonsen	Simonsen, near Quimby, Iowa	8,430 ± 540
Lindenmeier	South of Colorado-Wyoming boundary, northern Colorado	8,000 to 12,000
Wapanucket	Southeastern Massachusetts	9,000 to 11,000
Hell Gap	Near Guernsey, Wyoming	9,000 to 15,000
Cochise	Whitewater Draw, northwest of Douglas, southeastern Arizona	9,000 to 15,000
Fort Rock Cave	Fort Rock, Oregon	9,188 ± 480
Lubbock	Lubbock, Texas	9,883 ± 350
Arlington Springs	Santa Rosa Island, California	10,000 ± 200
Bullbrooks	Northeastern Massachusetts	10,000 to 11,000
Folsom	Near Folsom, New Mexico	9,500 to 10,500
Great Salt Lake	Great Salt Lake, Utah	10,000 to 15,000
Natchez	Natchez, Mississippi	10,000 +
Clovis-Portales	Between Clovis and Portales, central-eastern New Mexico	12,000 to 16,000
Ventana Cave	Castle Mountains, southern Arizona	8,000 to 11,000
Midland	Near Midland, west-central Texas	10,000 to 11,000
Sandia	Northern part of Sandia Mountains	11,000 to 13,000
Brewster	Eastern Wyoming	10,375 ± 700
Domebo	Near Stecker, Oklahoma	11,045 ± 647
Palouse River	Near Washtucna, Washington	12,000

Areas of Indian culture and
principal tribes in North America

MACKENZIE AREA: 1. Kutchin; 2. Ingalik; 3. Tanana; 4. Tanaina; 5. Han;
6. Dog Rib; 7. Yellow Knife; 8. Tahltan; 9. Slave; 10. Sekani; 11.
Beaver; 12. Chipewayan; 13. Tsimshian; 14. Sarsi; 15. Carrier

NORTHWEST COAST: 1. Tlingit; 2. Tsimshian; 3. Haida; 4. Kwakiutl;
5. Nootka; 6. Coast; 7. Salish; 8. Chinook; 9. Takelma; 10. Hupa;
11. Yurok

PLATEAU: 1. Thompson; 2. Okanagan; 3. Flathead; 4. Nez Perce; 5.
Bannock; 6. Snake; 7. Klamath

six miles. During winter it is actually possible to walk across the strait, and it is easily navigable by boat.

The people who entered the Americas by this route were mostly of Mongoloid origin, though it is possible that some non-Mongoloids may have entered, too. This last possibility is suggested by the fact that many Indians, even before admixture with whites, were known to exhibit predominantly Caucasoid characteristics.

There were probably several waves of migrations of different populations, some of whom made their way down to the very tip of South America, namely Tierra del Fuego. We know with certainty that men were already living in Tierra del Fuego at least eleven thousand years ago, and they were probably there much earlier.

Some of the sites in which human remains have been found either in the form of human bones or human artifacts are listed in Table 5 with the estimated age of these remains, established wherever possible by the radiocarbon method. However, it is to be noted that there is in fact no positive evidence of the presence of man in the New World earlier than fifteen thousand years ago.

KEY (*continued*)

SOUTHWESTERN: 1. Navaho; 2. Hopi; 3. Tewa; 4. Keresan; 5. Zuñi; 6. Havasupai; 7. Mohave; 8. Yuma; 9. Pima; 10. W. Apache; 11. Mescalero; 12. Seri; 13. Papago; 14. Tarahumara; 15. Yaqui; 16. Mayo

CALIFORNIA BASIN: 1. Klamath; 2. Modoc; 3. N. Shoshoni; 4. Paviotso; 5. Shoshoni; 6. Karok; 7. Miwok; 8. Pomo; 9. Mono; 10. Yokuts; 11. Salinan; 12. Southern Paiute; 13. Ute; 14. Mohave

PLAINS: 1. Blackfoot; 2. Assiniboine; 3. Gros Ventre; 4. Flathead; 5. Crow; 6. Bannock; 7. Mandan; 8. Hidatsa; 9. Arikara; 10. Dakota; 11. Ponca; 12. Omaha; 13. Osage; 14. N. Shoshoni; 15. Arapaho; 16. Cheyenne; 17. Kiowa; 18. Wichita; 19. Comanche; 20. Mescalero; 21. Caddo; 22. Tonkawa; 23. Lipan

NORTHEASTERN: 1. Naskapi; 2. Cree; 3. Montagnais; 4. Beothok; 5. Assiniboine; 6. Ojibwa; 7. Ottawa; 8. Dakota; 9. Fox; 10. Sauk; 11. Menomini; 12. Winnebago; 13. Kickapoo; 14. Iowa; 15. Illinois; 16. Miami; 17. Algonquin; 18. Huron; 19. Iroquois; 20. Erie; 21. Micmac; 22. Abenaki; 23. Penobscot; 24. Massachusetts; 25. Lenape; 26. Powhatan

SOUTHEASTERN: 1. Shawnee; 2. Cherokee; 3. Yuchi; 4. Chickasaw; 5. Creek; 6. Lipan; 7. Tonkawa; 8. Natchez; 9. Choctaw; 10. Seminole; 11. Timucua

MEXICAN: 1. Tepehuane; 2. Huichol; 3. Cora; 4. Otomi; 5. Aztec; 6. Tarascan; 7. Zapotecan; 8. Maya

Areas of Indian culture and principal
tribes in South America

THE RADIOCARBON METHOD OF DATING

It was possible to say two paragraphs ago that we know with certainty that man was living in Tierra del Fuego at least eleven thousand years ago because in recent years an accurate method of dating ancient remains has become available. This is the radiocarbon method.

The radiocarbon method of dating ancient remains depends upon the fact that radioactive carbon (carbon 14), which is liberated in the atmosphere as a result of the interaction of cosmic rays with nitrogen, is present in the structure of all living matter. During their lives all living things maintain a constant percentage of carbon 14 in their carbon structure. This percentage is the same for every form of life. At death the assimilation of carbon and carbon 14 ceases, and the carbon 14 atoms begin to disintegrate. Disintegration of carbon 14 atoms proceeds at a constant rate, 15.3 atoms per minute per gram of carbon. The rate of disintegration of carbon 14 was checked on samples of known age, yielding, in general, the most satisfactorily accurate results. The studies of Professor W. F. Libby and his colleagues conducted at the University of Chicago have shown that after 5760 ± 30 years have elapsed, one half of the carbon 14 atoms have disintegrated. After about 11,136 years only a quarter of the atoms will be left, and so on. The radiocarbon method will yield reliable results as far back as 70,000 years.

KEY TO MAP ON FACING PAGE

ANTILLEAN: 1. Carib
CHIBCHAN: 1. Cuna; 2. Chibcha
ANDEAN: 1. Quitu; 2. Quechua; 3. Aymara; 4. Atacama; 5. Calchaqui
AMAZONIAN: 1. Bare; 2. Arawak; 3. Wapisiana; 4. Tama; 5. Macusis; 6. Moxos; 7. Witoto; 8. Boro; 9. Jivaro; 10. Manaos; 11. Apiaka; 12. Piro; 13. Chacabo; 14. Bororo; 15. Tupi; 16. Tupinamba; 17. Kaingua
ARAUCANIAN: 1. Mapuche
SOUTHWESTERN: 1. Chono; 2. Alacalaf; 3. Yahgan
ONAN: 1. Ona
PATAGONIAN: 1. Pehuenche; 2. Puelche; 3. Tehuelche
CHACO: 1. Choroti; 2. Mataco; 3. Toba; 4. Pilaga; 5. Macori; 6. Abipone
EAST BRAZILIAN: 1. Canella; 2. Cayapo; 3. Kaingang; 4. Bakairi; 5. Charante; 6. Botocudo; 7. Cayapo

THE POTASSIUM-ARGON METHOD OF DATING

By this method the age of organic materials is determined by the ratio of potassium 40 to argon 40 they contain. Potassium 40, a radioactive form of the element, is present in all living things. It decays into argon 40 at the rate of 50 per cent every 1.3 billion years. By determining the amount of potassium 40 that has decayed into argon 40, the passage of time since the specimen was part of a living system can be determined, since potassium is incorporated into tissues only during life.

≥ 6 ≤

THE EVOLUTIONARY FACTORS
IN THE DIFFERENTIATION
OF MAN

WE HAVE SEEN that in the prehistoric period there were different types of men. Today in a land such as the United States we see types of men drawn from the most diverse populations from many parts of the world. There are American Indians, the original inhabitants of the Americas, there are whites of different origins, Mongoloids, Polynesians, Negroes, and Asiatic Indians. The members of all these groups are fundamentally alike and superficially different. How did all those differences which physically distinguish the different ethnic groups from one another come into being?

The answer, so far as we know it, constitutes a fascinating story.

UNITY WITHOUT UNIFORMITY

Most authorities agree that all men have originated from a common ancestral stock. What that stock was, and when the diversification commenced, we do not know. But whatever the stock was we can be sure that the diversification did not proceed in straight lines, that whatever ethnic groups originated from that stock did not pursue their development independently of all the other ethnic groups. On the contrary, as we have already pointed out, such groups almost certainly met with others and interbred with them, a process which was undoubtedly often repeated. There is, therefore, an essential unity to mankind without any uniformity.

All men belong to the same genus *Homo* and the same species *sapiens.* Of this species there are many *varieties,* a word well chosen because it indicates that the different ethnic groups of men merely represent the variations on a common theme oriented toward the achievement of that great harmony of humanity, which so many of them seem so bent on destroying with their false shibboleths of nationalism and racism, and similar disorders of the mind. Just as the different instruments of an orchestra are necessary to the production of great music, so the various peoples of the earth are necessary to the harmony of the world.

It will help us to understand this better if we consider the nature of the factors which have produced the differentiations that we recognize in the varieties or ethnic groups of man.

THE FACTORS PRODUCING HUMAN DIFFERENTIATION

Beginning as a single human population early in the history of man, some families must have wandered off and set up separate domiciles at considerable distances from each other; and so in the course of time these isolated groups became populations in their own right, and upon them the following factors exerted their influence in producing evolutionary change: (1) natural selection, (2) mutation, (3) isolation, (4) genetic drift, (5) hybridization, (6) sexual selection, and (7) social selection.

NATURAL SELECTION

On the fifth page of his epoch-making book *The Origin of Species* Charles Darwin defined natural selection in the following words:

As many more individuals of each species are born than can possibly survive; and as, consequently, there is a frequently recurring struggle for existence, it follows that any being, if it vary however slightly in any manner profitable to itself, under the complex and sometimes varying conditions of life, will have a better chance of surviving, and thus be *naturally selected.* From the strong principle of inheritance, any selected variety will tend to propagate its new and modified form.

Today we speak of Darwinian fitness or adaptive fitness, meaning that such variations that arise and are beneficial to an organism tend, through the action of the environment, to be preserved and to preserve the organism, so that such organisms are more likely to leave a larger progeny than organisms lacking similar adaptive traits (Darwinian fitness). For the latter reason a good short statement of natural selection is provided by the phrase "differential fertility." Those who possess adaptively valuable qualities in their particular environment will be at an advantage in comparison with those who do not possess such qualities, and the former are likely to flourish and the latter not to do so well. In short, natural selection is selection primarily for reproductive efficiency.

Most traits, if not all, possessed by man are of adaptive value.

It might be supposed that the adaptive value of such traits as skin color, hair form, and nose shape, to mention several of the most obvious ones, would by this late date be fully understood. But this is not the case. The truth is that we do not know what the adaptive value of most such traits may be. There have been many theories, but few of them have withstood critical examination. Skin color is a possible exception. It is known that a darkly pigmented skin is an advantage in areas of high sunlight intensity and under conditions of prolonged solar radiation. Skin cancer in India, for example, is almost exclusively confined to Europeans, and cancer of the face among sailors is comparatively frequent among whites and rare among nonwhites. Severe exposure to solar radiation destroys the cell processes and even the cells themselves which carry the melanin pigment, or when the pigment cells, the melanocytes, are not themselves destroyed, blocks the transport of the pigment to the skin cells. Clearly, then, the adaptive value of darkly pigmented skin is to serve as a protection against radiation-induced damage.

The number of melanocytes are the same in different ethnic groups, but differ in their concentration of melanin from individual to individual.

Different environments call for different adaptive traits, and the different traits which human beings exhibit are probably due principally to the fact that man has had to adapt himself to whatever environments he found himself in. As the most adaptable of all creatures, it is not surprising that man should exhibit such a remarkably interesting variety of adaptive traits. Indeed, it may be said of man that his principal trait is his adaptability.

MUTATION

A mutation is a transmissible change in the structure of a hereditary particle or gene, the giant nucleic molecule whose biochemical components constitute the biological bases of heredity. Mutations are the raw materials of the evolutionary process. Without mutation there can be no significant evolutionary change. Mutations are constantly occurring in all populations. It is believed that at least one mutation occurs in every human being sometime between conception and death.

The kinky hair of Negroes may be due to a mutation having adaptive value, but what that adaptive value may be we do not know. All the anthropoid apes have straight hair. The inference is that the original ancestral group of man had straight hair, but that subsequent mutations in the ancestry of the Negroid group gave rise to kinky hair. Sometimes kinky hair appears in a white family as a mutation. There are several such families recorded in Scandinavia in which there can be no question of any but white ancestry.

Mutations having adaptive value would rapidly become established in the small populations which were characteristic of early man.

ISOLATION

The earth is a very large place, and upon its surface men have lived in small groups, until the very recent period, isolated from one another for long periods of time. Isolation means here the separation of a group from all other groups of the same species, so that breeding takes place completely or largely within the isolated group alone. Isolation is brought about by such natural factors as distance, mountain ranges, rivers, forests, seas, and the like.

Under conditions of isolation and random mutation—all mutation being random—every isolated group, no matter how like other groups it originally was, would more or less rapidly become distinguishable from all other groups by virtue of the differences in the mutations which would arise in each group.

This fact makes isolation a factor inseparable from the phenomenon of genetic drift.

GENETIC DRIFT

This phenomenon, first described by Professor Sewall Wright of the University of Chicago, is sometimes called "the Sewall Wright effect." It has already been pointed out that the populations of man up to the Neolithic or New Stone Age were very small. For example, two eminent authorities, Professors Herbert J. Fleure and Grahame Clark, have independently estimated that the total population of Britain (now 50 million) in Upper Paleolithic times (under a subglacial climate) was between 250 and 2000, probably nearer 250 in winter months. At the present day peoples in the lower hunter stage of cultural development, such as the Australian aborigines and the Eskimos, seldom attain a population size of over 400.

In such small populations, owing to the sampling nature of the hereditary mechanism, it is possible for mutations having no particular adaptive value to survive and spread throughout the particular population. Only time and isolation are required, and the random variation presented by mutation will turn up a number of neutral mutations which may become established in the population. What these are likely to be no one could possibly predict because they are random in their appearance. Such mutations due to chance fluctuations may increase or become extinguished.

HYBRIDIZATION

This has, in the past, been a much-neglected factor in considering the differentiation of man. Hybridization simply means the mating of individuals differing from one another in one or more genes or traits. There are two kinds of hybridization: (1) between individuals and (2) between populations. Both types of hybridization have played important roles in the evolution of mankind.

During hybridization, and as a net effect of hybridization, both the groups involved bring to each other their adaptively valuable biological traits and thus serve to compensate for any deficiencies which the other group may be carrying. Hence deficiencies which are brought to the mating process are generally canceled out. Hybridization has, therefore, a powerful creative effect. Not only does it produce the modification of physical (structural) traits, but it produces also the modification of physiological (functional) ones, and thus serves to increase the adaptive value of the offspring. As a result of such matings, there is generally an increase in size, fecundity, resistance to disease, and other adaptive qualities—just as is the case in hybrid corn or hybrid animals. This phenomenon is known as *hybrid vigor* or *heterosis*.

Human hybrids tend to show a greater resistance to disease. For example, in Tierra del Fuego the unmixed natives succumbed to the measles which the hybrids were able to resist. The evidence suggests that in man hybrid vigor (heterosis) expresses itself not so much in structural traits but by changes in fertility connected with subtle mechanisms of adaptive value such as, for example, potential resistance to disease. Differences in fertility within the same populations may also be due to the slight heterotic effects of many individual genes which in combination serve to produce such differences in fertility. It is also probable that levels of intelligence are not only prevented from declining as a result of hybridization but also refreshed, as it were, and enhanced by the introduction of new genes. It is also significant that congenital abnormalities are very much less frequent among the hybrid than among the non-hybrid members of any population.

SEXUAL SELECTION

It is not certain what part sexual selection has played in the past in the evolution of man. Sexual selection means the selection by the most powerfully endowed males of the most preferred females. "Powerfully endowed" may mean many different things. In one group it may mean the possession of the greatest muscular power, in another the greatest social prestige as a result of the ownership of property, in yet another it may mean being the kindest person, and in still another being the best hunter, the most loved, and so on. Such men may be in a favored position to choose what women they will marry or mate with. It is probable that to some extent sexual selection in this sense has been operative in the past in the evolution of man.

In fact it would seem that sexual selection has played rather more of a role in the recent past—and may well become an increasingly more important one in the future—than it has in the remote past. For not only are women being selected by males for their attractive qualities, but women are beginning to exercise their free choice for the first time on a large scale.

The preference of males for fat women in some societies tends to favor the increase of such body types in these societies. In America, mating between whites and Negroes and the preference of darker Negroes for lighter Negro females tends to lighten, in the long run, Negro skin color. Among whites, the preference of brunets of one sex for blondes of the other, and vice versa, constitutes an example of the manner in which a balanced distribution of such types may be maintained in a society.

SOCIAL SELECTION

In all populations social distinctions are instituted between individuals or groups of individuals. There are chieftains, members of the council, medicine men, sorcerers, accomplished hunters, great braves, members of the royal family, the nobility, the upper classes, the middle classes, the lower classes, and so on.

In nonliterate societies marriage does not proceed at random but is usually strictly regulated. Precise regulations often determine the breeding structure of the population in such a way that distinct differences may be produced between different segments of the population. In our own time we have certain outstanding examples of this: the Hapsburg jaw characteristic of many members of the Spanish royal family; bleeders' disease, or hemophilia, exhibited by a proportion of the male descendants of Queen Victoria, in which royal lady's father the mutation for this disorder most probably occurred; the insanity that afflicted many members of the house of Hanover. Intermarriage with women of good stock has resulted in the virtual elimination of the latter unfortunate affliction in the descendants of this house, who have occupied the throne of England for the last hundred years.

In earlier times social selection probably played its part in the differentiation of populations and of groups within those populations. If by social selection we understand the regulation of breeding between socially distinguished individuals or groups

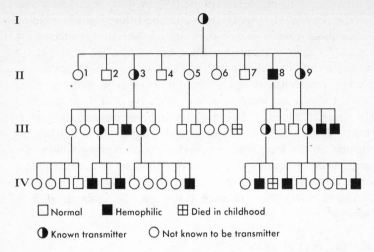

| □ Normal | ■ Hemophilic | ⊞ Died in childhood |
| ◐ Known transmitter | ○ Not known to be transmitter |

Hemophilic pedigree originating with Queen
Victoria. Circle = female, square = male

within a population, so that mating occurs between individuals
preferred by such social standards, rather than at random, we
may predict that social selection will continue to play a part in
the differentiation of man, both within populations and between
populations.

CULTURE

The effect of the man-made part of the environment, culture,
upon the physical evolution of man has, until recent years, been
strangely neglected. Yet culture has, undoubtedly, played a
most important role in the evolution not only of man's behavioral traits but also of his physical ones. For example, as a
consequence of the development of tools many functions formerly performed by teeth were more efficiently accomplished by
these artificially manufactured implements. Large teeth became
unnecessary; hence, they were gradually reduced in size, as
were the jaws which gave support to them. In hunting a great
deal of body heat is generated. This is most effectively dissi-

pated by the development of a large number of sweat glands and the loss of body hair which tends to retain heat. Both these changes have occurred in man. Thick hair is retained only in regions where it serves a protective function, as on the head, and at delicate marginal areas such as the eyebrow region, and in areas in which chafing might otherwise occur, as in the armpits.

The necessity for a large warehouse in which to store the essential information acquired by man as a member of the human culture contributed to an increase in the volume of his brain and to its complexity.

In adaptation to the hunting way of life man developed the erect posture, and this necessitated the development of large buttocks, principally made up of the gluteus maximus muscles, which are the chief muscles in maintaining the upright posture. These muscles serve to pillow the bony pelvis, in adaptation, among other things, to man's novel habit of sleeping on the ground.

The foot of man has developed in adaptation to stalking, walking, and running, which are all activities involved in the hunting way of life.

As a consequence of toolmaking and tool usage the hand has become a magnificent precision instrument.

THE PROCESS OF DIFFERENTIATION

The preceding discussion actually constitutes an account of the manner in which the so-called races of man have come into being. The term "race" has been confused by so much emotion and false meaning that it is better not to use it at all in the case of man. The noncommittal term or phrase "ethnic group" is to be preferred. An ethnic group is any population which as a group is distinguishable from another population on the basis of its inherited physical traits.

In the course of the physical differentiation of man, man has also had to adapt himself culturally to his many different environments, that is to say, he has had to work out a way of life best fitted to the particular environment in which he found himself. The particular way in which a people adapts itself to its environment behaviorally is called its *culture*. There has been a cultural differentiation of man as well as a physical one, and cultural factors have played an important role in the physical evolution of man. Because cultures vary, and often vary with a difference of ethnic group, some people have erroneously concluded that culture and "race" are biologically linked. This is to commit the fallacy of thinking that because two things are associated, one is the cause of the other. When conditions are associated we still have to look for their cause, and the cause of their association.

THE MENTAL UNITY OF MAN AND THE VARIETY OF HUMAN CULTURES

The outstanding characteristic of man is that he is the most educable, the most intellectually malleable, the most plastic of all creatures. Everything he knows as a human being, man has had to learn from other human beings. The evidence, as scientists have been able to reveal it, indicates that the average person in any human society is able to learn just as much as the average person in any other society.

Insofar as behavior is concerned, evolution has not proceeded to differentiate human populations into specialists able to meet the requirements of a particular environment and no others, but, on the contrary, the evolution of man has proceeded in such a manner that he has become a creature capable of adapting to *all* environments.

Indeed, as we study human beings in all societies we find that the one trait which is at a premium in human society is *adaptability,* plasticity, the ability to get along with others, to adjust to rapidly changing conditions. It is not any special ability which is emphasized, but rather this general ability of adaptability, plasticity. And we may be sure that it has been so throughout the history of man's sojourn on this planet. If this is so, then it constitutes a strong argument for the mental unity of mankind on an evolutionary basis, and an argument for the cultural differentiation of humanity on precisely the same basis. Human beings everywhere can learn to do the things that other human beings have anywhere done.

Now that the airplane has put the most distant place on the earth no more than a day away, the tendency of human populations is the opposite of what it was in the prehistoric period, when small populations tended to migrate away from a common center toward the periphery of the earth. Today human populations are tending to come together physically. It is our immediate task, and it will be the realization of the future, to enable the peoples of the earth to come together culturally and spiritually. Then all mankind will once again become truly one—maintaining, it is to be hoped, the differences as far as it is desirable to maintain them, and producing a genuinely creative, interanimating unity without a dulling uniformity.

THE ETHNIC GROUPS
OF MAN

IF WE WANT TO THINK and talk clearly about the different groups of mankind—and we shall increasingly have to do so as time goes on—it is desirable to have some scheme by which we can distinguish one group from another. Most classifications of the ethnic groups—the so-called races—of mankind are based on the recognition of a similarity within each group of physical traits based on heredity.

There are three major groups of mankind: (1) the Negroid, (2) the Mongoloid, and (3) the Caucasoid. A major group is a complex of populations which are classified together because of a basic similarity of certain physical traits.

THE NEGROID MAJOR GROUP

This group is characterized by dark brown skin, often almost ebony-black, and in some populations yellowish brown. The body hair is very sparse; the head hair varies from tightly curled to peppercorn (sparsely distributed tufts). The nose tends to be broad and rather flattish, the lips usually thick and everted, and there is some forward projection of the jaw. The ears tend to be small, the head long.

There are three subgroups: (1) African Negroids, (2) Oceanic Negroids, of the Territory of New Guinea and the great group of islands extending to the east all the way to the Fiji Islands, and (3) Negroids of southeastern Asia, including the Andamanese of the Bay of Bengal, the Semangs of the Malay Peninsula and East Sumatra, and the Aetas of the Philippines.

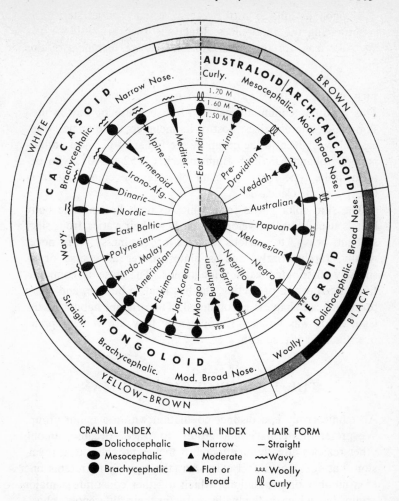

CRANIAL INDEX	NASAL INDEX	HAIR FORM
⬤ Dolichocephalic	► Narrow	— Straight
● Mesocephalic	▲ Moderate	～ Wavy
● Brachycephalic	▲ Flat or	₷₷₷ Woolly
	Broad	ℓℓ Curly

Diagrammatic representation of the major and
ethnic groups of man

THE MONGOLOID MAJOR GROUP

The skin sometimes has a faintly yellowish tinge, and a fold of
skin overhanging the inner angle of the eye opening—the epi-
canthic fold or Mongoloid fold—is present in most individuals.
Body hair is very sparse, and head hair is black, lank, and fewer
to the square centimeter than in Caucasoids.

Mongoloids inhabit northern, central, and southeastern Asia —embracing the Philippines, Malaysia, the East Indies—and the Americas, for the American Indians are members of this major group.

THE CAUCASOID MAJOR GROUP
This group is commonly called white because of the "white" skin which characterizes most of its members. The name "Caucasoid" was given to the group by the father of physical anthropology, Johann Friedrich Blumenbach (1752–1840), who described the type from the skull of a female emanating from Georgia in the Caucasus.

The skin color varies from white to dark brown. The body hair is usually well developed, and head hair varies from silky straight, wavy, to various degrees of curliness. It is never woolly, rarely frizzly, and is almost never as coarse or as sparsely distributed as in Mongoloids. The nose is comparatively narrow and projecting, the lips tend to be thin, and every form of head shape is encountered. The original habitat of the Caucasoids was Greater Europe; now they are widely spread over the face of the earth.

THE ETHNIC GROUPS OF MAN

An ethnic group is a distinct population within a major group. We prefer the term "ethnic group" to "race" because, among other reasons already mentioned, it makes no claim to a precision that does not in fact exist. An ethnic group represents one of a number of populations, which together constitute a major group, but which individually maintain their differences, physical and cultural, by means of isolating mechanisms, such as geographic or social barriers. These differences will vary as the powers of the geographic and social barriers vary. Where these barriers are of low power, neighboring ethnic groups will intergrade or hybridize with one another. Where these barriers are of high power, such ethnic groups will tend to remain distinct from or replace each other.

A classification of the major and ethnic groups of man is presented on pages 108–115.

TABLE 6. AVERAGE CRANIAL CAPACITIES IN ANTHROPOIDS AND MAN. MALES.

Cubic Capacity	Group	Method	Cubic Capacity	Group	Method
97	Gibbon	m	1440	Combe Capelle	e
125	Siamang	m	1442	South American Indians	c
400	Chimpanzee	m	1446	Modern Europeans	c
416	Orangutan	m	1450	Neanderthal	mc
506	Gorilla	m	1450	Ingwavuma	e
506	Plesianthropus	e	1451	Polynesians	m
562	Australopithecus africanus	e	1456	Sandwich Islands	m
530	Zinjanthropus boisei	e	*1457*	Swanscombe	c
500	Paranthropus robustus	e	1460	"Kaffirs"	m
680	Homo habilis	mc	1467	Chinese	m
775	Homo erectus II	e	1468	Swiss	m
900	Homo erectus robustus	mc	1470	Fontéchevade	mc
940	Homo erectus I	e	1473	Western Eskimo	m
1000	Homo erectus (Olduvai)	e	1475	Japanese	m
1043	Sinanthropus	mc	*1474*	Bury St. Edmunds	e
1100	Solo	mc	1476	Maoris	c
1177	Steinheim	c	1480	Ehringsdorf	e
1300	Saccopastore I	c	1480	Modern English	m
1230	Cape Flats	mw	1487	Tahitian	m
1243	Baining	c	1490	Buriats	m
1250	Saldanha	c	1490	Koreans	m
1256	Australian aborigines NT	c	1495	E. Centr. Amerinds	c
1264	Tasmanians	c	1498	Kalmucks	m
1264	Andamanese	m	1500	Springbok Flats	e
1278	Australian aborigines S A	c	1501	Ancient Europeans	c
1280	New Guinea	c	1505	Upper Paleolithic	c
1285	Veddahs	m	1516	Central Eskimo	c
1305	Rhodesian man	mc	1519	Iroquois	c
1307	La Quina	c	1525	Spy I	c
1300	Talgai	c	1530	Chancelade	c
1320	Saccopastore II	c	1532	Algonquin	c
1323	Melanesians	m	1532	Matjes River	c
1329	Bushman	mc	*1540*	Galilee	e
1333	Gibraltar adult	mc	1540	Tepexpan	mw
1335	Hindu and Tamil	m	1550	Monte Circeo	c
1338	Australian aborigines Victoria	m	1552	Mount Carmel	c
1346	African Negroes	m	1552	Early Neanderthals	mc
1350	American Negroes	mw	1552	Classic Neanderthals	mc
1359	Tyrolese	m	1564	Le Moustier	c
1383	Ainu	mc	1573	Mongols	m
1386	London (Lloyds)	e	1575	Singa	c
1388	Galla and Somali	c	1590	Předmost	c
1403	Dayaks	c	1593	Keilor	m
1406	Burmese	m	1600	Fish Hoek	c
1408	Hottentots	c	1600	Teshik-Tash child	e
1415	Aetas	c	1610	Shanidar I	m
1425	Spy II	c	1625	La Chapelle-aux-Saints	mc
1425	Châtelperron	mc	1641	La Ferrassie	mc
1424	Malayans	m	1650	Wadjak	c
1427	Marquesans	m	1650	Gibraltar boy	e
1434	Moriori	c	1660	Cro-Magnon	c
1435	1st to 2nd Dynasty Egyptians	c	1680	Elementeita	c
1438	Czechs	m	1700	Boskop	e
1439	Fijians and Loyalty Islanders	m	*1925*	Zitzikama	c

Capacities in *italics* indicate that 10 per cent has been added to the capacity of the female skull which is the presumed sex from which this particular type is known. Female capacities may generally be approximated by deducting 10 per cent from the male capacity. In the human species the cranial capacity is about 200 cc more than the volume of the brain.

m = measured with seed; mc = measured and calculated; mw = measured with water; c = calculated; e = estimated

THE MAJOR AND ETHNIC GROUPS OF MAN

Major Group: NEGROID

African Negroes
Ethnic Groups:
   a. The True Negro: West Africa, Cameroons, and Congo
   b. The Half-Hamites: East Africa and East Central Africa
   c. Forest Negro: Equatorial and Tropical Africa
   d. "Bantu-Speaking Negroids": Central and Southern Africa
   e. Nilotic Negro: Eastern Sudan and Upper Nile Valley
   f. Bushman: Southern Angola and Northwest Africa
   g. Hottentot: South Africa

Oceanic Negroids
Ethnic Groups:
   a. Papuans: New Guinea
   b. Melanesians: Melanesia

African Pygmies or Negrillos
Ethnic Group:
   a. African Pygmies or Negrillos: Equatorial Africa

Asiatic Pygmies or Negritos
Ethnic Groups:
   a. Andamanese: Andaman Islands
   b. Semang: Central region of Malay Peninsula, and East Sumatra
   c. Philippine Negritos: Philippine Islands

Oceanic Pygmies or Negritos
Ethnic Group:
   a. New Guinea Pygmies: New Guinea

Major Group: CAUCASOID

Ethnic Groups:
   a. Basic Mediterranean: Borderlands of the Mediterranean Basin
   b. Atlanto-Mediterranean: Middle East, Eastern Balkans, Africa, Portugal, British Isles
   c. Irano-Afghan Mediterranean: Iran, Afghanistan, parts of India, Arabia, and North Africa

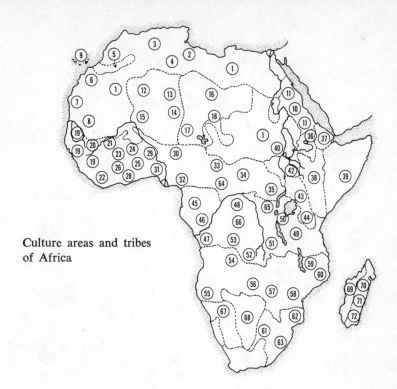

Culture areas and tribes
of Africa

KEY TO MAP

NORTH AFRICA AND MAURETANIA: 1. Arabs; 2. Berbers; 3. Kabyles; 4.
Chamba; 5. Chleu; 6. Tekna; 7. Ouled Delim; 8. Brakna; 9. Guanche;
10. Bicharin; 11. Bedja

SAHARA AND LIBYA: 12. Tuareg Kel Ahaggar; 13. Tuareg Kel Ajjer; 14.
Tuareg kel Aïr; 15. Tuareg Ifoghas; 16. Tibu; 17. Daza; 18. Teda

SUDANESE: 19. Fulbe, Peul; 20. Malinke; 21. Bambara; 22. Temne; 23.
Senufo; 24. Dogon; 25. Mossi; 26. Lobi; 27. Agni; 28. Ashanti; 29.
Songhi; 30. Hausa; 31. Yoruba; 32. Ibo; 33. Sara; 34. Banda; 35.
Zande

ABYSSINIAN: 36. Amhara; 37. Danakil; 38. Galla; 39. Somali

NILOTIC: 40. Nuba; 41. Dinka; 42. Nuer; 43. Turkana; 44. Masai

BANTU: 45. Fan; 46. Teke; 47. Kongo; 48. Mongo; 49. Nyamwezi; 50.
Nkole; 51. Luba; 52. Lunda; 53. Kuba; 54. Tchokwe; 55. Ambo;
56. Rotse; 57. Tonga; 58. Roswi; 59. Yao; 60. Makua; 61. Tswana;
62. Thonga; 63. Zulu

PYGMIES: 64. Mbenga; 65. Mbuti; 66. Tswa

KHOISAN: 67. Hottentot; 68. Bushman

MADAGASCAR: 69. Sakalava; 70. Merina; 71. Betsileo; 72. Bara

Culture areas and tribes of Asia

KEY TO MAP

SIBERIAN: 1. Eskimos; 2. Chukchee; 3. Koryak; 4. Yukaghir; 5. Ainu;
6. Yakut; 7. Lamut; 8. Samoyede; 9. Dolgan; 10. Evenki-Tungus;
11. Vogul; 12. Ket; 13. Ostyak; 14. Gilyak; 15. Goldi; 16. Oroch

NORTHERN EUROPE: 17. Votyak; 18. Cheremis; 19. Chuvash; 20. Mord-
vin; 21. Bachkir; 22. Lapps; 23. Zyrian

CENTRAL ASIA: 24. Russians; 25. Kalmuk; 26. Kazak; 27. Karakalpak;
28. Turkoman; 29. Uzbek; 30. Tajik; 31. Kirghiz; 32. Iigur; 33.
Hakka; 34. Karagasi; 35. Tuba; 36. Altai Oyrat; 37. Buriat; 38.
Khalka (Mongols)

ORIENTAL: 39. Japanese; 40. Koreans; 41. Chinese; 42. Tibetans; 43.
Mia-tse; 44. Dioi; 45. Nung; 46. Tai

SOUTHEAST ASIATIC: 47. Annamites; 48. Moi; 49. Khmer; 50. Thai; 51.
Korean; 52. Burman; 53. Khasi; 54. Katchin; 55. Lisu; 56. Moso;
57. Lolo

d. Nordic: Central Europe, Scandinavia and neighboring regions
e. East Baltic: East Baltic regions
f. Lapps: Northern Scandinavia, Kola Peninsula
g. Alpine: France along the Alps to Russia
h. Dinaric: Eastern Alps from Switzerland to Albania, Asia Minor, and Syria
i. Armenoids: Asia Minor
j. Hamites: North and East Africa
k. Indo-Dravidians: India and Ceylon
l. Polynesians: Polynesia (Central Pacific)

Sub-Group: AUSTRALOID or ARCHAIC CAUCASOID

Ethnic Groups:
a. Australian: Australia
b. Veddah: Ceylon
c. Pre-Dravidian: India
d. Ainu: Japan, Hokkaido (Yezo), and Sakhalin Islands

Major Group: MONGOLOID

Classical Mongoloids
Ethnic Groups:
a. An undetermined number of ethnic groups in the older populations of Tibet, Mongolia, China, Korea, Japan, and Siberia, including such tribes as the Buriats east and west of Lake Baikal, the Koryak of northern Siberia, the Gilyak of northernmost Sakhalin and the mainland north of the Amur estuary (who appear to have mixed with the Ainu), and the Goldi on the Lower Amur and Ussuri

KEY (*continued*)

FORMOSAN: 58. Tayal (Atayal); 59. Ami; 60. Paiwan
PHILIPPINES, MALAYSIAN, INDONESIAN: 61. Igorot; 62. Negritos; 63. Manobo; 64. Subanun; 65. Toradja; 66. Dusun; 67. Murut; 68. Punan; 69. Iban; 70. Dayak; 71. Semang; 72. Senoi; 73. Malays; 74. Batak; 75. Minangkabau; 76. Lampong
INDIAN: 77. Ona; 78. Veddah; 79. Singhalese; 80. Nayar; 81. Tamil; 82. Kanara; 83. Telgu; 84. Maratha; 85. Gond; 86. Munda; 87. Bihar; 88. Hindi; 89. Gujrati; 90. Bhil; 91. Sindhi; 92. Sikh
MIDDLE EASTERN: 93. Kurds; 94. Lures; 95. Persians (Iranians); 96. Bactrians; 97. Berbers; 98. Kafir; 99. Afghans; 100. Brahui; 101. Baluch

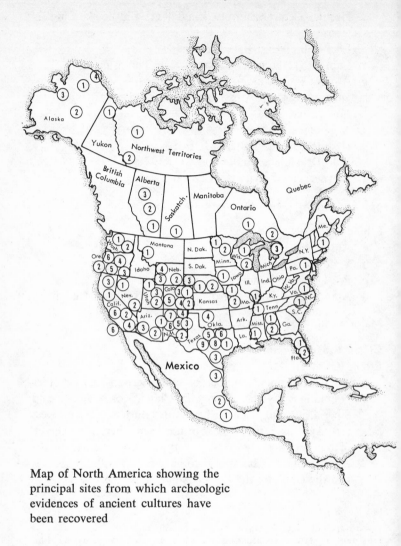

Map of North America showing the
principal sites from which archeologic
evidences of ancient cultures have
been recovered

KEY TO MAP

CANADA
ALBERTA: 1. Cereal; 2. Edmonton; 3. Peace River
SASKATCHEWAN: 1. Mortlach
ONTARIO: 1. The Brohm Site; 2. George Lake; 3. Manitoulin Island
NORTHWEST TERRITORIES: 1. Great Bear Lake; 2. The Pointed Mountain
    Site
YUKON: 1. The Engigstciak Site

KEY (*continued*)

MEXICO: 1. Tepexpan; 1. Santa Isabel Iztapan; 1. Tequixquiac; 2. El Horno; 3. Tamaulipas
UNITED STATES
ALABAMA: 1. The Quad Site; 2. Russell Cave
ALASKA: 1. The Brooks Range; 2. The Campus Site (Fairbanks); 3. The Iyatayet Site (Cape Denbigh); 4. The Point Barrow Sites
ARIZONA: 1. Cochise Sites; 2. The Lehner Site; 2. The Naco Site; 3. Ventana Cave
CALIFORNIA: 1. The Borax Lake Site; 2. Lake Mohave; 3. Lower Klamath Lake (The Narrows); 4. The Pinto Basin; 5. Santa Rosa Island; 6. The Stahl Site (Little Lake)
COLORADO: 1. The Claypool Site; 2. The Dent Site; 3. The Lindenmeier Site; 4. The Linger Site; 5. The Uncompahgre Plateau; 4. The Zapata Site
FLORIDA: 1. Melbourne; 2. Vero
ILLINOIS: 1. The Modoc Rock Shelter
IOWA: 1. Turin
KENTUCKY: 1. The Parrish Site
MASSACHUSETTS: 1. The Bull Brook Site
MINNESOTA: 1. The Browns Valley Man Locality; 2. The Minnesota Man Locality
MISSISSIPPI: 1. The Natchez Pelvis Locality
MISSOURI: 1. Graham Cave; 2. The Nebo Hill Sites
MONTANA: 1. The MacHaffie Site
NEBRASKA: 1. The Allan Site; 2. The Meserve Site; 1. The Lime Creek Site; 1. The Red Smoke Site; 3. The Scottsbluff Bison Quarry
NEVADA: 1. Fishbone Cave; 2. Gypsum Cave; 1. Leonard Rock Shelter; 2. Tule Springs
NEW MEXICO: 1. Bat Cave; 2. Burnet Cave; 3. The Clovis-Portales Area; 4. The Folsom Site; 5. The Lucy Site; 6. Manzano Cave; 3. The Milnesand Site; 7. The Rio Grande Sites; 8. The San Jon Site; 6. Sandia Cave; 1. The Wet Leggett Site
NORTH CAROLINA: 1. The Hardaway Site
OREGON: 1. Catlow Cave No. 1; 2. Fort Rock Cave; 3. The Klamath Lake Area; 4. Odell Lake; 5. The Paisley Caves; 6. Wikiup Dam Site No. 1
PENNSYLVANIA: 1. The Shoop Site
SOUTH DAKOTA: 1. The Ray Long Site (Angostura Basin)
TEXAS: 1. Abilene; 2. Colorado (Lone Wolf Creek); 3. Freisenhahn Cave; 4. The Lipscomb Site; 5. The Lubbock Site; 6. The McClean Site; 7. Miami; 8. The Plainview Site; 9. The Scharbauer Site (Midland)
UTAH: 1. Danger Cave; 2. Black Rock Cave; 2. Dead Man Cave; 3. Promontory Cave
VERMONT: 1. The Reagan Site
VIRGINIA: 1. The Williamson Site
WASHINGTON: 1. Lind Coulee; 2. Palouse River
WISCONSIN: 1. Effigy Mound; 2. Hopewell
WYOMING: 1. Agate Basin; 2. The Jimmy Allan Site; 3. The Finley Site; 4. The Horner Site

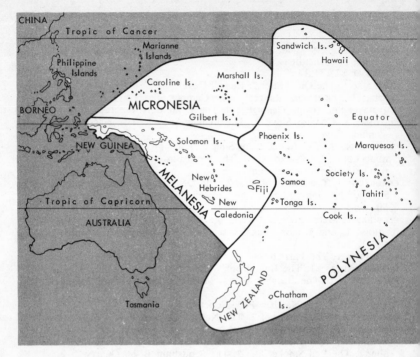

Divisions of Oceania

Arctic Mongoloids
Ethnic Groups:
  a. Eskimo: Extreme northeast of Asia, Arctic coast of North America, Greenland. The type includes the Aleuts of the Aleutian Islands, and the Reindeer and coastal Chukchee of northeastern Siberia
  b. Evenki or true Tungus (Americanoids): Mongolia, Siberia, Asiatic highlands north of the Himalayas
  c. Kamchadales: Kamchatka
  d. Samoyedes: Kola Peninsula, White Sea and Yenisei regions

The Mongoloids of the extreme northeast of the Asiatic continent are distinguished as the *Paleoasiatics*. These are considered to be the complex of ancient populations of Asia who migrated early to this extreme peripheral region. The populations believed to have migrated later into the northeast of the Asiatic continent are known as the *Neoasiatics*.

Paleoasiatics
Chukchee, Koryak, Kamchadale, Gilyak, Eskimo, Aleut, Yukaghir, Chuvantzi, Ostyak of Yenisei, Ainu

Neoasiatics
Finnic tribes, Samoyedic tribes, Turkic including Yakut, Mongolic, Tungusic

American Indians
Ethnic Group:
   a. An undetermined number of ethnic groups of North, Middle, Central, and South America

Indo-Malay
Ethnic Groups:
   a. Indonesian: Southern China, Indo-China, Burma, Thailand, interior of Malay Archipelago
   b. Malay: In addition to Indonesian distribution, Malay Peninsula, Dutch East Indies, Philippines, Okinawa, and adjacent islands

A WORD IN CONCLUSION

There have been many classifications of the "races" of mankind. The above is but one of them. With slight modifications, no doubt, it is one that would be accepted by most authorities. Nevertheless, it should be remembered that all classifications are arbitrary, and this one is as much subject to modification as others have been. The important thing is to avoid becoming the captives of our own classifications. The above classification is calculated to be helpful in thinking about the varieties of mankind; it is *not* intended to solidify anyone's views.

❧ 8 ❧

NEEDS AND
CULTURE

Every human being is born with certain basic needs, inborn needs which must be satisfied if the organism is to survive. The basic needs are the needs for: oxygen, food, liquid, rest, activity, sleep, bowel and bladder elimination, escape from frightening situations, and the avoidance of pain.

In connection with each of these basic needs, every human being largely is subjected to the teaching of his culture. We all breathe, eat, drink, rest, sleep, and eliminate in ways that we have learned are the custom in our group, no matter what the custom may be in other groups. We are all to a certain extent custom-made, tailored according to the prevailing pattern of our society.

THE MEANING OF CULTURE

Because of man's great capacity for adaptability and his remarkable ingenuity, he can improve in a great variety of ways upon the manner in which other animals meet their needs. Man has the ability to create his own environment, instead of, as in the case of other animals, being forced to submit to the environments in which he finds himself. Within every society there are particular ways in which needs are met. The origins of these ways are usually "lost in the mists of antiquity" or, as the Australian aborigines say, they "belong in the dream time." In fact, one of the hardest things in anthropology is to trace the origin of a custom. There is usually no one old enough to re-

member its origin because as a rule it came into being a very
long time ago, long before oral tradition or written history.

Culture represents man's response to his basic needs. Culture
is man's way of making himself comfortable in the world. It is
the behavior he has learned as a member of society. Culture may
be defined then as the way of life of a people, the environment
which a group of human beings usually occupying a common territory have created in the form of ideas, institutions, pots and
pans, language, tools, services, and sentiments.

It is this man-made environment, culture, which all human
societies impose upon the physical environment and in which all
human beings are trained. Culture becomes so identified with
life itself that it may fairly be said that it is not so much superimposed on life as that it is an enlargement of it. Just as a tool
enlarges and extends the capacities of the hand, so culture en-
hances and extends the capacities of life.

The criteria by which culture may be recognized are: (1) it
must be invented, (2) it must be transmitted from generation to
generation, and (3) it must be perpetuated in its original or in
modified form. While some other animals are capable of limited
cultural behavior and have a way of life that is learned, man alone
seems to be virtually unlimited in his capacity for culture. The
process of creating, transmitting, and maintaining the past in the
present is culture—the capacity which the American semanticist
Alfred Korzybski called time-binding. Plants bind chemicals, animals bind space, but man alone binds time.

Culture is the joint creation of the individual and society,
mutually and reciprocally interacting to serve, maintain, support, and develop one another.

Culture, then, is the complex of mental configurations which
in the form of behavioral and material products constitutes
man's principal means of adapting to the total environment,
controlling it, changing it, and transmitting and perpetuating
the accumulated means of doing so.

Evolutionary change in all animals proceeds by mutation and
the storage of the adaptively valuable mutations in the genes. In
man evolutionary change has also proceeded in this way, but
with the addition of a great many nongenetic changes which
also represent social evolutionary changes. These nongenetic
behavioral or cultural changes are stored not in genes, but in the

man-made, the learned, part of the environment, in the culture, in the tools, customs, institutions, laws, ballads, and the like, and in the memories of men, as well as in other extragenetic devices for the storage and retrieval of information.

Human nature is what one learns from the man-made environment; it is not something with which one is born. What one is born with are the potentialities for learning which, by suitable teaching, are capable of being transformed into the uniquely human abilities.

THE CONDITIONS OF HUMAN CULTURE

Man is human by virtue of the fact that he possesses certain potentialities which are developed to a uniquely high degree. He is born utterly dependent for his survival upon other human beings during his first few years, and in this dependency relationship, more prolonged than that of any other creature, he learns to develop those potentialities under the stimulating influence of a human environment. *Whatever we know and do as human beings, we have had to learn from other human beings.* In comparison, other creatures have to learn very little. For much of what they do, they are equipped by instinct, that is, by automatic action without reasoning or design. Man differs from all other creatures in the possession of the following characteristics, which at the same time constitute the conditions for the development of human culture:

1. Freedom from the instinctive, automatic reactions to environment that characterize much animal behavior.
2. Extraordinary plastic potentialities for the development of a complex intelligence. Educability.
3. A highly developed capacity for symbolic thought.
4. Speech.
5. A highly developed capacity for innovation.

The creature endowed with these qualities is in a position to interact in a creative manner with its environment—and that, precisely, is what human beings have done from the beginning of time.

THE DEVELOPMENT OF CULTURE

When a man and a woman unite and produce a child, the biological family comes into being. The helplessness of the child renders necessary the expenditure of much time and energy in its care. Men and women who have united in such a social relationship will have to provide the infant with special comforts and protection. Domestic activities will have to be put on a new basis. The infant will have to be fed, be cleaned, have his needs attended to, and be trained. Each parent will have to assume his or her special obligations and share somewhat different kinds of authority. The arrival of the child entails the development of a new relationship with the members of neighboring families and the social (legal) recognition of the new bonds which have been established between the parents and the child. In turn, recognition of the new bonds which have been created by the parents with the group must be established, for the parents have now become responsible to the group for the proper rearing and education of the child.

Such obligatory relationships will inevitably arise in all societies under the conditions described, and the economic, legal, educational, and political responses—that is, the responses growing out of the satisfaction of the basic needs and of the new needs derived from the ways in which they are satisfied—in turn give rise to all or almost all of those cultural responses which we know. Most of these cultural responses may be fruitfully summarized in the following synopsis:

CULTURAL RESPONSES TO THE NEEDS OF
HUMAN BEINGS LIVING IN SOCIETY
1. Patterns of communication: gestures, language, writing, etc.
2. Material traits
   a. Food habits and food-getting
   b. Personal care and dress
   c. Shelter
   d. Utensils, tools, etc.
   e. Weapons

    f. Occupations and industries
    g. Transportation and travel
 3. Exchange of goods and services: barter, trade, commerce
 4. Forms of property: real and personal
 5. Sex and family patterns
    a. Marriage and divorce
    b. Methods of reckoning relationships
    c. Guardianship
    d. Inheritance
 6. Societal controls
    a. Mores
    b. Public opinion
 7. Government
    a. Political forms
    b. Judicial and legal procedures
 8. Religious and magical practices
 9. Mythology and philosophy
10. Science
11. Art: carving, painting, drawing, dancing, music, literature, etc.
12. Recreational interests: sports, games, etc.

CULTURE AND THE INDIVIDUAL

No single individual ever gains a knowledge of the whole of his culture. As a member of his culture, the individual is equipped to participate in it, not to become a mere repository of it. Every individual is born with a unique biological endowment of potentialities which are like those of his fellows, but not exactly the same. This is the biological heredity of the individual. The culture into which the person is born constitutes his social heredity. The interaction between the individual's biological and social heredities is, in fact, what constitutes the person's *heredity*. There is no heredity without the interaction between one's biological equipment of potentialities and the environment or environments in which they undergo development. A man's nature is not what he is born with, but what he becomes under the organizing influence of the socializing environment into which he is born.

Thus human nature is largely the expression of human nurture—the product of the interaction of genetic potentialities with the culturalizing factors which work upon them to give their general nature its particular form.

It is principally through the agency of the stimulation of the cultural environment that the individual becomes a person.

THE COOPERATIVE NATURE OF MAN

A human baby is born not only wanting and needing to be loved, but also wanting and needing to love others. For too long we have erroneously believed that babies are born rather selfish, aggressive little creatures, who need to be disciplined and repressed. This belief has done a great deal of damage to human beings and to society. The evidence indicates that all human beings are born with their needs oriented in the direction of love. The word "love" is used here to mean behavior which confers upon others survival benefits in a creatively enlarging manner.

Every society exhibits this desire of human beings to live together in peace and creative harmony. Many societies also exhibit the evidences of competition, sometimes to a most violent degree. Such evidences constitute the symptoms of disoperation rather than of health. The general trend of human evolution in most human societies has been toward the achievement of cooperation between men rather than of conflict. By "cooperation" is meant striving together to achieve common goals. By "competition" is meant striving against others to achieve a goal. Cooperative competition, when it is in the interest of the group, can be healthy, but competitive competition for selfish interests is not.

An example of cooperative competition is any sport that we play in such a manner as to bring out the best in the other players so that they in turn will be caused to bring out the best in us. In competitive competition one takes every possible advantage of the opposite side in order to win at any cost, rather than rejoicing in the fact that one has done one's skillful best and that the better team has won, whether it be one's own or the opposite side's.

In recent years several writers have published books in which they have maintained that man is an innately aggressive creature, driven to all sorts of violent behavior by his "instincts" of aggression and territoriality, a "naked ape" whose "raw animal nature" makes him the intractable creature they consider him to be.

These ideas are easily disposed of. In the first place, man has no instincts. In the course of man's evolution on the savannas, where he learned to take to hunting and where the highest premium was placed upon cooperation, problem solving, and adaptability, biologically predetermined reactions, that is, instincts, would have served no useful purpose. Instincts are not adapted to meet anything but the conditions for which they were evolved. New challenges of the environment require novel *responses,* thought-out problem-solving responses, not automatic reactions. The appropriately successful response to the particular challenge of the situation, that is, the use of intelligence, is what is required, and it is because man has evolved in a man-made environment which necessitates the ability to make such responses that he has become the intelligent creature that he is.

Hence, if man has no instincts such aggressiveness as he sometimes displays must be otherwise accounted for. Here, too, the answer to the question as to the origin of man's aggressiveness is simple: Man learns to be aggressive. Men are not born with aggressive drives, that is to say, with predispositions to inflict injury upon others. Rather, they learn to behave aggressively from the models of aggression in which they are conditioned in their particular environments. There are whole societies such as those of the Pueblo Indians, the Tarahumara Indians of Mexico, the Ifaluk of the Pacific, the Bushman of Southwest Africa, the Pygmies of Central Africa, Australian aborigines, Eskimos, and others, that are characterized by their unaggressiveness. It has been argued that these peoples have simply learned to overlay their innate tendency to aggression with unaggresiveness. The answer to that argument is that there is no evidence whatever that they have done so or that there exists anything remotely resembling inborn aggressiveness, except possibly in some chromosomally abnormal individuals.

There is no more evidence for an "instinct of aggression" than there is for an "instinct of territoriality." Many animals do tend to hold and defend what they come to regard as their particular territory, but few monkeys and none of the great apes indulge in such behavior. Whatever could have happened to the "instinct of territoriality" in such creatures? Perhaps the simplest explanation is that they never had such an "instinct" any more than man does. The truth is that different societies behave in very different ways about territory. Some are devoted to their territories and jealously defend their boundaries. Others, like the Eskimo, have no sense of territoriality and welcome all comers. Hunting and food-gathering peoples often live on lands the boundaries of which overlap and never constitute the cause of conflict of any kind. There are other societies that peacefully adapt themselves to those who encroach upon their lands and often move on elsewhere. Still others experience no difficulty in abandoning their homelands for others more favorable to their purposes.

In short, some societies are not territorial minded while others are. This has nothing whatever to do with instinct and everything with what these peoples have learned to feel and think about territory.

Similarly, when some writers speak of overcrowding as eliciting aggressive reactions, it is not that the aggression thus evoked is reactive, but responsive; not innate, but learned. Crowding as such will not elicit aggression. Asiatic Indians, Hong Kong Chinese, Indo-Chinese, the Dutch, the Swiss, the Pueblo Indians constitute varying examples of crowding and overcrowding characterized by an almost entire absence of aggression.

❧ 9 ❧

ATTEMPTING TO MAKE
THE BEST OF IT

Human beings are social creatures; they take pleasure in association, they become depressed in the absence of their fellows. Hence, one always considers man as a social being, and as such we find that the first experience to which he is usually exposed is the sound of the human voice. Even before he is offered nourishment, the human baby is greeted by his mother's voice or the voices of those who are assisting in his birth. The sound of the human voice is an experience for the human infant more constant than the experience of food.

Small infants learn to understand human sounds much earlier than was at one time believed possible, even though they do not begin to speak much before they are fourteen months old. What and how a human being speaks is learned largely from the members of his immediate cultural group. At the present time there are more than three thousand languages in the world. Of these many languages, the vast majority fall into about thirty-five unrelated linguistic stocks or families.

COMMUNICATION

Communication is the process of receiving and transmitting information, signals, or messages, whether by gesture, voice, or other means. It is the unit of the social situation.

Language may be defined as any *system* of communication between individuals. *Speech* is the verbal form of language which conveys information by chopping vocal tones into pulsations of discrete vibrating segments. Man alone can articulate such speech segments, and by means of his speech organs transmit symbols to others.

Language is the formal code, the institution, the abstraction. Speech is the uttered message, the vocal act. We *speak* a language according to the agreed, formal rules. The languages of men are mutually translatable, the languages of animals are not—they are at most interpretable.

Sometimes anthropologists are asked whether it is true that primitive languages do not have a grammar. The answer is that if there were any system of sounds that was grammarless, whatever else it might be, it certainly wouldn't be a language. Grammar comprises the formal rules by which the meanings of sounds are governed. The fact is that many "primitive" languages aren't any more primitive than most of the rest of the culture; indeed, they are often a great deal more complex and more efficient than the languages of the so-called higher civilizations.

As with all other things, man has been most inventive in his patterns of communication.

It is important to distinguish between communication, language, and speech. These terms may, of course, be used synonymously, but more precisely communication refers to the transmission or reception of a message, while language, which is usually used interchangeably with speech, is here taken to mean the speech of a population viewed as an objective entity, whether reduced to writing or in any other form.

Thoughts and feelings may be communicated in nonverbal ways, as through movements of the body. This is the study of *kinesics*. Kinesic communication is a highly efficient form of nonverbal communication.

Communication, then, includes all those processes by which people influence one another, speech being the most important of all the special ways of influencing human beings.

We know practically nothing of the history of any language. The earliest written language, Sumerian cuneiform, goes back to about 3500 B.C. In comparison to the total history of man this represents less than 2½ minutes of the human day.

Speech comes into being when two or more individuals agree to attach the same meanings to the same sounds, and thereafter to use those sounds consistently with the meanings that have been conferred upon them. The meanings are symbols which stand for something which is actually not present to one's senses in physical form. Language or speech, then, is behavior made artificially clear. It is a tool.

THEORIES OF THE ORIGIN OF SPEECH

Speech is thought and feeling made explicit. Concerning the origin of speech there have been many theories, and most of them have been abandoned by serious scholars. Some of these theories go by rather amusing names and are worth mentioning. They are as follows:

1. The *bow-wow* theory: that speech is imitative of sounds, such as the barking of dogs and other animals, etc.
2. The *pooh-pooh* theory: that speech is derived from instinctive ejaculations evoked by pain or other intense feelings or sensations.
3. The *dingdong* theory: namely that there is a mystic harmony between sound and sense, that to every impression within there is an appropriate expression without—"everything which is struck rings. Each substance has its peculiar ring."
4. The *yo-he-ho* theory: that words first came into being as a result of the expressions following upon strong muscular effort. It would relieve the body to let the breath come out strongly and repeatedly, to "heave" and to "haul."
5. The *gestural* theory: that speech came into being when men began to imitate with their tongues the gestures they made with their bodies.
6. The *tarara-boom-de-ay* theory: that speech originated in the collective half-musical expressions of early man when his principal vocal exercise was a meaningless humming or singing. Suppose that a group of such early men have together succeeded in bringing down a great mammoth, they will at once spontaneously be filled with joy and will strike up a chant of triumph, say something like "tarara-boom-de-ay." The sounds of the chant might easily come to mean, "We have brought him down. We have conquered. Hooray! Let us give thanks." And so on from there. This theory has been suggested by Professor Otto

Jespersen, and it is remarkably true that what cannot be spoken can often be sung.

7. The *tally-ho* theory: that speech came into being as a result of the necessities of communication arising during the hunt (see below).

Perhaps none of these theories provides an entirely satisfactory explanation of the origin of speech, but there can be little doubt that some of the mechanisms suggested by each of them have entered into the formation of every spoken language, and possibly constituted elements in the origin of every form of speech.

Speech is the communication of ideas by articulate sounds or words. Speech is the means by which the biological is transformed into the cultural dimension. Speech is the medium through which the biological creature is changed into a culturally active member of his society, the means, indeed, by which the culture is transmitted.

Man is the only creature who, by natural selection, became, and continues to be, increasingly and progressively more dependent, first upon toolmaking, and second, upon linguistic communication.

THE HUNTING OR "TALLY-HO" THEORY
OF THE ORIGIN OF SPEECH

Man's nearest relatives, the great apes, are all capable of vocalizations of various sorts, varying from cries and whines to squeaks, screams, growls, and roars. The orangutan and the gorilla tend to be quiet and taciturn. In the natural state the chimpanzee, although occasionally boisterous, also tends to be quiet, although in captivity he can be very noisy. Anatomically the great apes are equipped with all the vocal arrangements necessary for speech, but they do not speak. Speech has no adaptive value under the conditions of life in the forest, and there is therefore no evolutionary pressure to develop it. Human beings take about two years to learn how to talk and about seventy how to keep their mouths shut. The apes have done better than that.

> When manlike apes began to walk
> They took the step that led to talk,

And ably learning to walk erectly,
They never learned to talk circumspectly.

These ape-men, pressured by the chase,
Used howls and hoots to win the race,
And striving upward, changed their lot
To wordy man, the polyglot.

However that may be, when from a predominantly vegetarian life in the forest, the earliest progenitors of man were forced to take to a hunting way of life on the open plains, a high premium would have been placed upon the ability to communicate by voice with one's fellow-hunters. This would be especially true during the chase, when it would be highly desirable to communicate one's changing intentions in adaptation to the changing conditions during the hunt. This would require signaling of modified strategies and directions, and this could best be achieved by voice.

In the early history of man's development problem-solving abilities would have been at a high premium, especially during the hunt. Taken together with the ability to interpret and imitate the meaningful cries of one's companions, such a combination of abilities would have enjoyed high survival value. Thus the development of intelligence and of speech would have gone hand in hand. Those individuals having the ability to solve problems quickly would be more likely to leave a progeny than those wanting in such a combination of abilities. Therefore speech as well as intelligence would possess the highest adaptive value.

In stalking and in pursuing prey speech would be an invaluable aid. It would, however, appear unlikely that speech first developed during the stalking phase of man's economic activities, for in stalking prey it is essential to remain quiet, and under such conditions speech would not have developed. It is more likely that the stage of evolution during which speech originated coincided with that phase of man's economic activities in which he developed from a small game collector to an active hunter. Small game collecting appears to have been the way of life of the earliest men, the australopithecines. It is more likely that the earliest hunters would have attempted to make meaningful cries during the chase than during the collecting or simple stalking of game. In running, and during the excitement of the chase, there would be a strong tendency toward the violent expression of air from the lungs, an expression which could readily have been converted into a cry or yell. Such expressive cries or yells could ultimately be given definite meanings. Such meanings conferred upon them in the context of the situation in which they arose could readily be interpreted and imitated and thus established as meaningful sounds or words. In this manner a simple vocabulary would come into being, which could then serve as a basis for further development.

A vocal sound having a relatively stable meaning, and used as a unit of communication, is known as a *morpheme*. A morpheme is the smallest structural unit that has lexical or grammatic meaning. For example, the sound "s" at the end of a word is a morpheme, having the meaning "plural." Some birds can chop vocal sounds into discrete segments, but they do not communicate thoughts by such means, for it is man alone who produces cerebrated symbols by means of speech.

From the very birth of speech the meaning of a word has been the action it produced, and that is still the meaning of a word. Speech, born in a cooperative context for a cooperative purpose, is the greatest ligature that binds human beings together. The natural purpose of words is to put man in touch with his fellow man. As Henry James remarked, "All life comes back to the question of our speech—the medium through which we communicate."

It is probable that speech was among the earliest, if not the earliest, of the tools originated by man. Because of its high adaptive value speech undoubtedly played a considerable role in the evolution of man from his earliest beginnings. That the faculty originated under the pressure of necessity, and was perpetuated by natural selection, is an assumption which is in conformity with all that we know concerning the evolution of many of man's other traits. Thus, to the inarticulateness of nature man added a new dimension, speech. It is speech that makes man human.

LANGUAGE, GRAMMAR, AND CULTURE

Every language has a definite phonetic system, that is, the sounds which are used are limited to a specific number of vowels and consonants. However, the actual sounds employed or neglected may differ very widely in different languages. Similarly, languages differ even more markedly in their grammars. There are grammatical processes in American Indian languages, for example, which are quite unknown to languages of the great Indo-Germanic stock. Such a process, for example, is that of infixing, which means taking a particle of a word and putting it in the middle of another word or between the particles of several other words, and giving the sentence a modified meaning by doing so. In English we have prefixes and suffixes but no real infixes, although the cockney's "abso-bloody-lutely" or "im-bloody-possible," which introduces a whole word into the middle of the original to give it greater intensity, is based on the principle. In sounds we have no clicks, but some languages like African Bantu are characterized by a well-developed system of meaningful clicks.

The Eskimos live among ice all their lives but have no single word for ice. This is not due to their stupidity, for they are very intelligent indeed, but rather to the fact that they have no need for such a word. What they do need are many words meaning the different states and conditions of ice: ice solid, ice melting, ice moving, ice breaking up, ice piling up, and so on. Their language reflects the practical uses to which it must be put. There are many nonliterate peoples who live in forested regions who have no word synonymous with our "tree." In this case, too, the language reflects practical need, and there are names for every kind of tree and every state of tree. Language reflects cultural interest, and for most peoples the meaning of a word is related to the action with which it is connected. As languages grow somewhat removed from practical need, they tend to become more abstract and unrelated to action. Language is marked by its great plasticity and its adaptability to the needs of the culture.

Where life is dependent upon sharply defined meanings, the language will reflect this. To most American Indians the statement "A dog is barking" sounds singularly unenlightening. What he wants to know is: *What* dog? *Whose* dog? *Where?* Is it standing, running, jumping, or what? In his own language he can say these things in as few sounds as we utter when stating that a dog is barking. It is important for the Indian to have the kind of information he needs, and he would never think of being as vague as we are in speaking of a barking dog. Unlike ourselves, many nonliterate people have the capacity of saying a maximum number of things in a minimum number of words.

Language molds thought to a much greater extent than is generally appreciated. Every language enshrines its own reality, for the world is organized according to the manner reflected in the language. Hence, the best means to the understanding of the way in which other people think is through their language. Different languages do not merely represent different ways of labeling experience, but each language serves as a guide to reality in that it is by means of its words and grammatical categories that order and meaning is given to the world of experience.

The analysis of language, *linguistics,* is complex and interesting but a vast subject, into which it would be impossible to enter any further here. We must conclude, then, with the statement that without language human culture would be impossible, and that granting man's great educability, he becomes even more educable through the exploitation of the new meanings that language, properly used, makes possible. Language is the great social stimulator and binder.

Some day, it is to be hoped, humanity will speak a single international language (without necessarily losing the advantages of the language of one's own culture), and thus will the greatest obstacle to the communication between peoples be at once removed.

GESTURE

Gesture has been described as an auxiliary language. Some peoples, such as the southeastern European Jews and the Italians, do, indeed, to some extent make use of gestures in this way, but there are many other peoples who either make no use of gestures at all—like the American Indian, who is traditionally known to be both laconic and undemonstrative, and like the English, who to some extent may be similarly characterized—or who make very little use of gesture, such as the Melanesians of the Pacific.

Some Plains Indian peoples did elaborate a small inventory of gestures by which they were able to communicate with

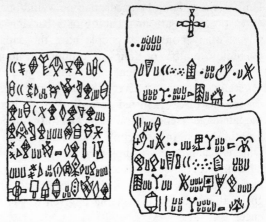

Earliest writing: Proto-Elamite

others, but these gestures were never a substitute for their spoken language, merely a supplement on occasion. There is no evidence that human language was ever preceded by a stage in which gestures alone were used for the purposes of communication.

WRITING

Writing is written language. It is the method of communicating thoughts and feelings by means of visible signs. The earliest writing that we know is cuneiform (Latin: *cuneus* = wedge, *forma* = form) from Sumer in Mesopotamia in the valley of the Tigris and Euphrates rivers—the very birthplace of civilization itself. The date is about 3500 B.C. So writing is just over five thousand years old. The so-called primitive peoples of the world have no written language and are therefore more appropriately spoken of as nonliterate. All peoples were nonliterate up to five thousand years ago. Egyptian hieroglyphics (sacred writings) are known from 3000 B.C. and are believed to have originated under the stimulus of Sumerian writing.

The forerunners of writing are known from cave drawings, rock drawings, message sticks, tree inscriptions, and bark letters. Below is reproduced a letter from an American Indian girl to her lover. It is a letter, but it is not writing because it is not a formalized system of symbols which everyone would understand; the marks are private, not public, symbols. The girl, who belongs to the bear totem, writes to her boyfriend, who belongs to the mudpuppy totem, that she would be glad to receive a visit from him, as indicated by the symbol of a hand drawn on the tepee which she occupies. The trail leads toward the lakes, shown by three irregular outlines, whence it branches off in the direction of the two tents, which are situated near those occupied by three Christian girls, indicated by the crosses.

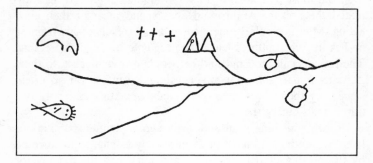

American Indian girl's letter

Many American Indian tribes had a form of pictographic writing, but with the exception of the Maya Indians of Yucatan, these forms of writing were very limited in function. The Maya Indians are the only aboriginal inhabitants of America who invented a system of writing. This hieroglyphic writing was finally deciphered in 1961 by three young Russian mathematicians with the aid of a computer. In 1821, after twelve years of work, Sequoyah, a Cherokee American Indian, singlehandedly invented a syllabary or alphabet of eighty-six characters which enabled the Cherokee Indians to become the second literate aboriginal people of the Americas. This invention constitutes an example of "stimulus diffusion."

OTHER FORMS OF COMMUNICATION

Forms of communication by "jungle telegraph" such as by drumbeat, smoke signals, whistling, horn blowing, cries, light, semaphore, sign language, and the like have been independently devised by many different peoples.

FOOD HABITS AND FOOD-GETTING

The anthropoid apes are herbivores, that is to say, they live on a plant diet, and they have the long gut of all herbivores. So has man. Man has inherited a herbivorous gastrointestinal tract, but he is everywhere an omnivore, eating everything that is edible— even the most poisonous plants and animals after having eliminated the poison. Anthropoid apes spend most of their time eating. Man is similarly inclined, but as a result of his training he has in some cultures learned to regulate his habits of eating. The Italians call the English "the people of five meals"; to them it seems remarkable that anyone should want to eat so often, since they themselves eat a very slight breakfast and only two other meals during the day.

Up to about nine thousand years ago all human populations lived by hunting and most of them also by fishing, supplemented by the picking of berries, fruits, and nuts, and the digging for roots and tubers. Perhaps the first division of labor between the sexes was that the male became the hunter and the female the food-gatherer.

There are many peoples at the present day who still earn their living in this way, for example, the Australian aborigines, the Eskimos, the Pygmies of central equatorial Africa, the Andaman Islanders, and the Onas of Tierra del Fuego, to mention but a few. Earning a livelihood by hunting and food-gathering means that populations must remain small and that they must continually wander over considerable areas of land in order to avoid exhausting both the plant and animal life of a limited region. Such small populations, being nomadic, build no permanent habitations, and are equipped only with the lightest and minimum number of tools. Their impermanent shelters are, therefore, likely to be simple, like the Australian windbreak, which is built of the lightest twigs and grass and constitutes a protection against wind and sun. There are no pots or pans. Cooking is done over a fire with the food held on a skewer or the whole animal dumped on the fire.

Should one desire to boil water, all one does is dig a hole of the desired depth in the ground, harden the sides, pour the water in, and then drop stones which have been heated in a fire into the water. This method of heating water, many hundreds of thousands of years after it was first invented, may have caused some genius to think of taking the hole out of the ground, as it were, and making it portable in the form of a stone vessel or pot.

Not all hunters were nomads. If one feeds mainly on fish, for example, one may easily lead a settled life. The salmon-catching Indians of British Columbia built permanent plank houses in permanent villages. Handicrafts were perfected and wealth created, and with it nobles, commoners, and slaves.

Hunting and fishing call for great resourcefulness. The Australian aborigines, for example, invented the boomerang, a unique hunting implement which will return to the thrower if it does not hit the target. They use spears, and spear-throwers which increase the power behind the spear. Since the Australian aboriginal of the interior lives in a desert or semi-desert region, almost anything is edible to him. Such an aboriginal will think nothing of catching a poisonous snake with his foot, picking it up, biting off its head, and then skinning and consuming it. Had we been brought up under similar conditions we would, no doubt, be happy to do the same.

The bow and arrow never reached Australia, but what early Edison invented it or where or when it was invented we do not know; it was probably invented somewhere in the Old World. It is found throughout Southeast Asia and the islands of the Pacific, in Africa, and in the Americas. The earliest evidence for the bow is from the Neolithic of Spain some ten thousand years ago, though it may very well have been invented in the Upper Paleolithic (Upper Old Stone Age).

Bolas, which have been found in Middle Paleolithic sites, are among the earliest hunting weapons of which we have knowledge (see p. 57).

Nets, snares, and traps are widely used among hunting peoples, as are decoys. From cave paintings and rock drawings, it is clear that our prehistoric ancestors often disguised themselves as the animals they hunted and then moved freely among the herd. American Indians did this when hunting buffalo and

Master of the Animals from the
Cave of Les Trois Frères

bison, Eskimos do this sometimes when hunting reindeer, and
the Bushmen of South Africa follow this practice when hunting
ostrich.

The ingenuity of hunting peoples in obtaining their food
bears striking testimony to the great intelligence of the human
species. It would take a very large book indeed to give an
account of all the methods, but perhaps we can best conclude
this section with an account of one of the cleverest traps, and
one of the simplest, that man has ever developed.

This trap is designed to catch guinea fowl. It is known from
Madagascar and many parts of Africa. It consists of a ring of
clay about six to eight inches across and about one inch high,
modeled on a flat rock. Some groundnuts are put in the circle.
The nuts are a little too big for the fowl to pick up, but they
keep on trying. Every time a fowl pecks at one and misses, its
beak comes down hard on the rock. Being persistent creatures,
they keep this up until their heads swell and they go blind. The
trap owner every day visits the brush near the trap and picks up
the disabled birds. When chickens are raised on a concrete floor
the same thing is likely to happen—as every poultry raiser
knows.

Hunting peoples have no domestic animals with the exception of the dog, and many of them don't even have a dog. Some hunting peoples have learned to put the dog to use in hunting: the Onas of Tierra del Fuego, for example, use their dogs to scent out the guanaco (a deer-sized camel) and bring it to bay. The Australian aborigines love their dogs as pets; the dogs serve also as scavengers to keep the encampment clean. It is probably this scavengering function which first brought man and dog together. Certainly by the Mesolithic, some fifteen thousand years ago, if not earlier, man had already domesticated the dog and soon went on to domesticate plants.

It is probable that the domestication of plants and animals occurred in response to the challenge of an increasingly arid environment following upon the last pluvial (wet) period. The resulting desiccation forced the development of new means of subsistence.

AGRICULTURE

Agriculture, the domestication or cultivation of plants for the purpose of human consumption, was probably independently discovered by different human groups at different times in different parts of the world. The discovery more often than not was probably made by women, for the simple reason that women are everywhere the foragers, the food-gatherers, the seekers of roots and seeds, and presumably had better opportunities for observing the relation between accidentally dropped seeds and their germination. Women were probably the first gardeners and agriculturalists. Planting, weeding, and harvesting is still a woman's job in most nonliterate societies. It is only when agriculture displaces hunting as an economic activity and becomes a major job that men take over.

The earliest evidence we have of the domestication of plants is from a people who occupied the upper levels of the Mount Carmel caves in the Wadi-en-Natuf in Palestine. The Natufians date back to about 7000 B.C., and they appear to have occupied their caves till about 5000 B.C. The importance of the Natufians for us lies in the fact that among their implements were found the earliest known examples of agricultural implements, namely sickles. These consisted of toothed-flint sickle blades set in a row into bone or antler handles. The sickle blades are characterized by a high polish such as that derived from the silica contained in the stems of cereal plants. While we have no direct evidence that the Natufians actually harvested cereal plants in this way, the probabilities are high that they did so. Although it is unnecessary to sickle wild cereals since upon ripening the heads automatically burst open and scatter the seed (the best way to gather such grain is by rubbing the heads between the palms of the hands or by striking them with a stick so that the grain falls into a receptacle), the mutant cereal plants are of the nonscattering variety and have to be sickled, then threshed at leisure. If they were not so treated they would rot, and since the heads would not automatically open, they would not reproduce themselves. Hence, it seems highly probable that they must have been artificially opened and the seed planted by man.

The Natufians had no domestic animals other than the dog and were without pottery, though they made stone vessels. Mortars and pestles of stone suggest that they ground the cereals they almost certainly cultivated.

The earliest Middle Eastern agricultural settlement known in which there is direct evidence of plant cultivation has been excavated at Jarmo in the foothills of southern Kurdistan in northeastern Iraq. Here two-row barley and emmer wheat were recovered, as well as sickles, mortars, and pestles, and weighted digging sticks. Here, too, there was no pottery, but stone vessels had been made, as well as adz- and ax-blades sharpened by grinding. Small flint implements known as microliths were found here as among the Natufians. The villagers of Jarmo lived in mud-walled houses and bred stock animals such as goats, sheep, pigs, and cattle—the earliest evidence of domesticated animals in the Middle East, and the very same creatures that form the basic stock animals to this day. The Jarmo settlement dates back to about 6500 B.C. The Natufians were a Mesolithic folk. The Jarmo villagers made progress from the Mesolithic toward the Neolithic but did not altogether attain a Neolithic culture.

Mount Carmel in Palestine and Jarmo in Kurdistan are both situated within the region known as "the Fertile Crescent," the vast semicircle of land that fringes the great deserts from Palestine to the Persian Gulf. To the north lie the foothills of Kurdistan, through which the turbulent Tigris and Euphrates run their course, one for fifteen hundred and the other for sev-

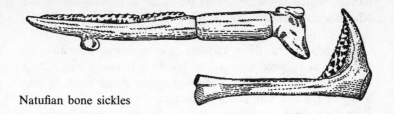

Natufian bone sickles

enteen hundred miles, emptying into the Persian Gulf. In this region conditions were especially favorable for the development of agriculture and the first permanent settlements.

In 1965 the discovery of the earliest known village site in Europe was announced. This site is at Nes Nikomedeia in the Macedonian plain of northern Greece. Along the shore of this early Neolithic settlement lived farmer-herdsmen who raised wheat, barley, and lentils, tended sheep and goats and almost certainly cattle and pigs. The radiocarbon date for the materials from this village is 6220 B.C. This date makes Nes Nikomedeia the oldest Neolithic site in Europe. Not only this, if the cattle and pig bones represent those of domesticated animals, and they almost certainly do—for most of the pig and cattle bones are those of immature animals, and such a large number of young

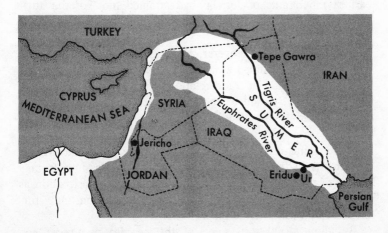

The Fertile Crescent

animals would not normally be the result of hunting—then Nes Nikomedeia represents the earliest dated occurrence of domesticated cattle anywhere in the world.

In addition to their farming and pastoral activities the Nikomedeians were engaged in hunting, fowling, and fishing. Deer, hare, and wild pig were among the game animals, and, living by the sea, fish, cockles, and freshwater mussels appear to have been staple articles of diet, as the presence of the bones and shells of these animals attests.

Clay figurines of both men and women which reveal a high development of artistic sensibility, with the female figurines outnumbering those of men, suggest the practice of fertility rites.

With the discovery of the new art of food production it was inevitable that supernatural beliefs and ritual practices would undergo further development and elaboration with the growing sophistication of the people.

In 1956 there was excavated at the base of Jericho, the city of Biblical fame, the earliest town thus far discovered. This Neolithic, pre-pottery Jericho was surrounded by a massive defensive wall, and contained stone houses with hog-brick floors, shrines, simple wooden furniture, and a good many other artifacts. The population, which may have numbered some two thousand, was devoted to a cult of skulls. It is inferred that the earliest inhabitants of Jericho had attained most of the attributes of civilization except those of pottery and writing almost nine thousand years ago.

According to its discoverer, Kathleen Kenyon, earliest Jericho probably resembled a medieval town, surrounded by fertile fields, and engaged in trade relations with other town dwellers living far away, to judge from the presence of such nonindigenous materials as obsidian, turquoise matrix, and cowrie shells.

If the claims made for Jericho prove to be sound, late village settlements in this region may represent retrogressions from a period of flourishing town life, due to increasing desiccation as conditions approached those of today in this region. Only further research will provide the answer.

By about 4000 B.C. some of the peoples of Lower and Middle Egypt, such as the Tasians, the Badarians, the Merim-

deans, and the Faiyumis, had made the passage into the New
Stone Age or Neolithic, the age characterized by the first
polished-stone and ground-edged implements, by the domestication of animals, by agriculture, and by the making of pottery.
The Tasians lived mainly by hunting and fishing and were
nomadic or seminomadic, but they had pottery and ground-
stone axes, and used rubbing stones for grinding small grains
like wheat or barley. The Tasians occupied a settlement at Tasa
near the east bank of the Nile in Middle Egypt.

The Merimdeans of Merimde Beni Sālame, a settlement on
the desert margin of the western edge of the Nile Delta, lived in
wood-framed huts in stockaded villages, kept cattle, sheep, and
goats, and cultivated cereals. The Faiyumis, who lived beside a
great lake (now desert) in the Faiyum in Lower Egypt, also
kept pigs. They hunted and fished and cultivated wheat and
barley which, like the Merimdeans, they gathered with sickles
and stored in large pits lined with straw matting. The bow and
arrow appears to have been the chief hunting implement.

These peoples, the Tasians, the Merimdeans, and the
Faiyumis, were Stone Age peoples, but the Badarians, who followed them, learned how to hammer copper, although they
never discovered how to melt, smelt, or mold it.

The pottery of these peoples we owe, no doubt, to the increased leisure which the cultivation of plants and the domestication of animals first made possible. Textiles were woven from
flax, other vegetable fibers, and wool, baskets from reeds, and
bags from various fibers. The earliest representation of a loom
is found on a pottery dish discovered in a woman's tomb at
Badari dating as far back as 4400 B.C.

The Badarians were followed by the Amratians, whose
appearance marks the beginning of the predynastic period in
Egypt and dates back to a little over 4000 B.C. The Amratians
were culturally much like the Badarians, but in addition made
use of alphabetlike signs—though not in a manner that could be
called writing.

Another culture that succeeded the Badarian, that of the
Gerzeans, who lived somewhat west of the Faiyum, reached a
high order of technical perfection in many respects. Its pressure-
flaked implements are among the most beautiful of their kind.
Copper, lead, and silver were known, and olives, in addition to

cereals, were cultivated. The rudiments of a script based on older Paleolithic hunting signs were in use, and the Gerzeans played a game resembling checkers. For the first time in the history of man, as far as we know, a king makes his appearance. It is believed that it was the Gerzeans who invented the calendar as an aid to agriculture, in order to foretell the date of the annual inundation of the Nile.

Following the Gerzean culture, there is a steady development in civilization: the overgrown villages become autonomous towns with their own officials; the use of copper and other metals becomes more frequent; the potter's wheel is discovered; and with the regular use of writing, history begins.

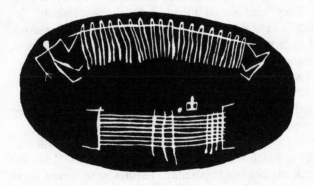

Loom depicted on early Egyptian pottery dish

STOCK-BREEDING AND PASTORALISM

It is interesting to note that the earliest domestication of animals should have occurred at approximately the same time as the cultivation of cereals. The principle underlying the two forms of food production is the same: the control of reproduction. As we have already mentioned, the earliest animals to be domesticated for the purposes of food are those we still use: cattle, sheep, goats, and pigs. Their bones are found in the earliest agricultural settlements of Jarmo in Kurdistan and in the settlements of the Faiyumis, Merimdeans, and Badarians in Neolithic Egypt. That the horse was used as a domestic animal in Elam, at the head of the Persian Gulf, earlier than 3500 B.C., is known from a representation of a man riding one. From Sumerian art of about 3000 B.C. we know that the wheeled cart was invented with horses and cattle as draft animals.

Pastoral communities, living largely off their herds, were well established by 4000 B.C., and there are many peoples today in Mongolia and Arabia who still follow this mode of subsistence, as do the Navahos of our own Southwest.

There appear to have been three major centers of origin for the domestication of animals: (1) The earliest, "the Fertile Crescent" of southwestern Asia, associated with seed farming and the domestication of herd animals, goats, sheep, cattle, and swine. Donkeys and zebu (humped) cattle were domesticated in areas bordering the Near East. (2) In southeastern Asia, associated with vegetative plantings and household animals, dogs, pigs, chickens, ducks, and geese. (3) An American center with (a) vegetative plantings in South America with domesticated llamas, alpacas, guinea pigs, and muscovy ducks, and (b) seed planting in Central America and domestication of the turkey.

PERSONAL CARE AND DRESS

Judging from existing nonliterate peoples, attention to the cleanliness of one's body and to one's appearance in general is a trait that occurs everywhere. In the matter of appearance it is principally the females, rather than the males, who excel, apparently because males everywhere place a high value upon physically attractive females.

It is an error to suppose that clothing is usually worn for protective purposes. In nonliterate societies it is worn mainly for decorative purposes. In most nonliterate societies, before the influence of the clothed white man, only such clothing was worn as was thought desirable, and what was thought desirable varied from society to society. In some societies it was considered immodest to expose certain parts of the body; in other societies complete nudity was the rule, as among the Australian aborigines and many African peoples. The Onas of Tierra del Fuego live in perhaps the worst climate in the world, a climate of bitter cold, snow and sleet, and heavy rains a great deal of the time, yet they usually remain entirely naked. During extremely cold

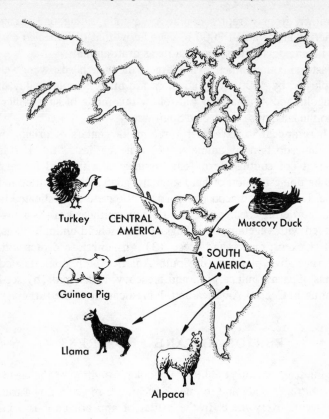

Probable centers of origin of major domesticated animals

weather they may wear a loose cape of fur and rub their bodies with grease.

The Australian aborigines live in a climate of daily extremes. During the day the temperature may be 120 degrees in the shade; at night it may fall below freezing! Yet they wear no clothes at all. At night the aboriginal builds a small fire and curls himself around it, waking every so often to put on a few more twigs in order to keep the fire burning.

Among the American Indians of earlier times everything from tailored clothes to complete nudity could be found. The

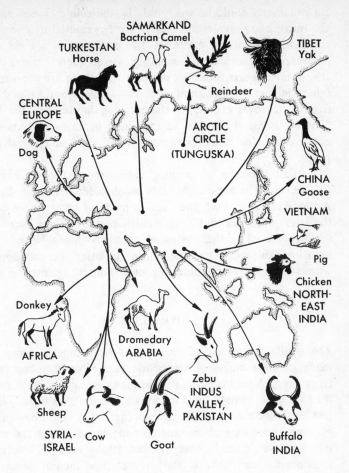

Indians in the vicinity of San Francisco Bay resorted to coats of mud to keep themselves warm. The Athapascan-speaking Indians of the Mackenzie River still make their garments of tanned hides. Eskimos, during the winter months of the year, are forced to clothe themselves in the warm furry skins of the animals they hunt. Their clothing constitutes, in fact, a masterpiece of ingenuity.

Clothing in some societies is used to symbolize differences in social status. Only great warriors could wear the feather bonnet that has come to be associated with the Plains Indians. Large ear ornaments could be worn only by the ruling classes among the Incas, and only certain classes were permitted to wear special feather-decorated garments. In our society a hospital doctor wears a doctor's coat, a working man frequently wears overalls, a nurse wears a nurse's uniform, and there are the various other kinds of distinguishing apparel—soldiers' and sailors' uniforms, the royal ermine of the reigning house, and the like. Clothing is clearly not to be explained in terms of biological needs alone, for it serves many psychological and social purposes.

Earliest evidence of clothing comes from the Neolithic of France, but it is almost certainly much older than that. In the Aurignacian period we find needles made of bone, as we do in the later Solutrean and Magdalenian periods. These were undoubtedly used in the sewing of garments. Since Neanderthal man lived during part of the Würm glaciation, he undoubtedly wore some form of clothing to protect himself against the cold.

SHELTER

The gorilla makes use of rock ledges from time to time, and many of the remains of prehistoric man have been recovered from caves. *Sinanthropus* (Peking man) is the earliest form of man we know to have lived in caves. Cave dwellers usually lived just inside the mouth of the cave, because they could have a fire burning a good deal of the time near the opening, whereas in the interior it would smoke everyone out. It may come as a surprise to some to learn that at the present time human beings, in millions, are still living in caves. And not only in Mongolia or Arabia, but also in Europe. At the present time there are over a million people in Spain alone who make their permanent habitation, as their ancestors have before them, in caves.

Where there are no caves, every people builds some form of shelter for protection, depending upon the available materials and the need. Nomads build impermanent shelters; sedentary peoples build substantial ones. But there are always exceptions. The Polar Eskimos are nomadic, but they build permanent houses of stone, to which they return after the summer's wanderings. The snowhouse, or igloo, is used only as a temporary shelter which they readily abandon when it is no longer needed. A movable tent of hide serves a similar purpose during the summer hunting season. The Onas of Tierra del Fuego, in an extremely rainy and cold climate, build only simple windbreaks of guanaco skins supported by poles to protect them from the weather. The Australian aboriginal builds a similar windbreak made of twigs and leaves, but on occasion he will erect a more substantial hut, of the simplest kind, of wood and grass. The Plains Indians, nomadic hunters, used the tepee, a skin tent which was easily moved. The Crow Indians erected a tepee made of many buffalo skins sewn together which was twenty-five feet high and could accommodate more than twenty people.

An elaborate portable shelter is the *yurt,* made by the Kazak of central Asia. This consists of a light wooden framework covered with felt. Some yurts have wooden doors and are separated into several rooms; the floor is beaten earth covered with rugs. The yurt is usually transported by oxen, horses, or camels, and can be erected in a half hour.

Clearly, the type of house or shelter a people erects will depend upon a number of factors, principally on the requirements of their environment and only secondarily on their technological development. The British Columbian fishing Indians, who built large, elaborate houses of wood that were permanent and immovable, are a special case since they were able to store enough food after the salmon run to last them a year and so were able to lead the life of wealthy "farmers."

The Iroquois of the eastern woodlands built "long houses," which were communal dwellings, rectangular in shape, some 20 to 30 feet in width, about 50 to 150 feet in length, and about 20 to 30 feet in height. From a central corridor running the full length, separate apartments opened on each side. Built of bark, long houses were substantial structures.

The Pueblo structures of New Mexico and Arizona are houses built of stone laid in adobe mortar or of adobe clay bricks that have been dried in the sun. The roofs, also of stone or adobe, are supported by log beams. These houses are communal dwellings, the separate apartments being built in stepped tiers four or five stories high. The skyscraper stone buildings of Lhasa, Tibet, were built long before our own.

Whether hut or duplex apartment, a house is something more

than a shelter. It is also a place where the family can be together, and where, together and alone, they can enjoy a certain amount of privacy. However gregarious a person may be, there are many times during his life when he enjoys the pleasures of solitude—absolute or relative. The privacy of the house affords the family and the individual an opportunity for reflection and careful planning.

WEAPONS

There is no evidence of the manufacture of weapons as instruments of offense or defense against other men until the Neolithic. When some groups began to cultivate herds of livestock, raiding by other groups soon followed. In order to protect property against marauders, defenses were erected, and new and more weapons were manufactured. Sometimes a marauder was captured, leading to the discovery that instead of killing the enemy one could enslave him. Thus was slavery born, in the Neolithic of what is now the Middle East.

A Neolithic raid may not have been comparable to a Second World War, but it was war nonetheless. After the horse was introduced into America, raiding for horses became a common practice among the Plains Indians, but the raid was usually carried out without any desire to harm the owners of the horses. Often as many as a hundred horses would be taken by the raiding parties, as well as mules and cattle. Not infrequently these raiding parties would result in bloody conflicts.

In the Neolithic the increase in the number of weapons found—battle-axes, rapiers, and swords of bronze or iron— suggests that either individuals or groups or both had made the sorry discovery that by acquiring the property of one's neighbor, one could increase one's own wealth, that war, in fact, seemed to be an economically productive activity—a totally erroneous belief which has bedeviled men down to the present day. It was in the Neolithic that man started off on the wrong foot with the discovery that the acquisition of large amounts of property leads to power, and that when one has power the only thing remaining to achieve is—more power. Food-gatherers and simple cultivators don't seem to think this way—they are too busy making a living—but pastoral peoples and mixed farmers seem to hanker after the acquisition of cattle and of slaves. Nonliterate people like the Australian aborigines and the Eskimos do not engage in warlike activities at all. In fact, it is extremely difficult to make them understand that peoples exist on earth who engage in such activities.

The raiding of some American Indians is to be regarded as a form of warfare, but the headhunting activities of New Guinea natives, and until recently the Dayaks of Borneo, are not properly so regarded. The motivation in such cases is not really infliction of injury upon another tribe in order to obtain its property, but rather to obtain for ritual purposes a magical increase in strength from one or more of its slaughtered members.

Weapons have been made of every conceivable material, from wood to tempered steel, and doubtless every possible peaceful implement that could be used for the purposes of war has been so used. Since wood is perishable, we know very little concerning its uses as material for weapons, but from the bone and stone points which have been recovered by the thousands, it is clear that in many cases they were associated with a wooden handle or shaft. Sling stones, used as ammunition, found in Iran date back to about 4500 B.C., mace-heads have been found at Merimde, and battle-axes become abundant during the Bronze Age (1500 B.C.).

Though some people, like the raiding Indians to whom we have already referred, certainly raided in order to acquire the property of their neighbors, Australian aborigines will take revenge on a member of another tribe for wife-stealing or for a previous killing of one of their own folk. Polynesians have attacked another group in order to obtain a victim for sacrificial killing, and so have the Ashantis of Africa and the Aztecs of Central America.

OCCUPATIONS AND INDUSTRIES

Occupational differences came into being with human society itself. The division of labor between the sexes was the first of these. Woman's work was of the domestic kind. In fact, it might be said that in most nonliterate societies woman was man's chief domesticated animal. Her work has always been considered menial. This is not to say that she is anywhere thoughtlessly treated or regarded, as it has sometimes been erroneously stated, as a beast of burden, but merely that man's work has been of a wider, and in general socially more esteemed kind. It is only in the most civilized societies that this way of regarding woman is being gradually rectified. In many lands of the Western world, the position of woman in society is being re-evaluated, especially the importance of her functions as a mother and homemaker, who is at the same time worthy of full equal political and social rights with men.

In Neolithic graves where one often finds dual burials of husband and wife the wife is usually much younger than the husband. Wives were probably buried alive with their husbands as a rule, but sometimes, as we know from the arrowheads found in their bones and around the skeleton, they were killed before the burial.

It has already been stated that women, as the gatherers of plant foods, a role they perform in every hunting and food-gathering society at the present day, were the inventors of horticulture and probably also of agriculture. The Neolithic revolution was therefore probably the creation of women, its development the work of both men and women. But we may be certain that the raiding activities which came in the Late Neolithic were the doing of men. It will be the task of women to reteach men how to conserve rather than to destroy.

In every nonliterate society there are medicine men, priests, elders of the council, skilled artists, skilled stoneworkers, skilled woodworkers, metalworkers, educators, engineers, and so on. As society advances there is an increase in the differentiation of occupations until many of them become institutionalized in the form of guilds, educational institutions, and legal institutions. An institution is an enduring, complex, integrated, organized behavior pattern through which social control is exerted and by means of which the fundamental social desires or needs are met. Institutions come about through the agreement of large numbers of people on the performance of certain functions, with each person still performing his occupational tasks separately, whether in medicine, law, the church, or education.

Industry came into being with human culture, and we have evidence of early industry in the form of the flint tools which early man made. It took great skill to manufacture these tools, and not everyone could make them. There is little doubt that in many early cultures gifted workers in stone were employed as specialists, and possibly the latter trained other individuals in the manufacture of tools. It is also quite possible that workers in stone taught other groups how to make tools in return for some consideration. There is reason to believe that the teaching of art techniques was transmitted on bone or stone (see p. 223), so that others many hundreds of miles away could ben-- efit.

In the great riverine depressions of the Nile valley, on the alluvial plains between the Tigris and the Euphrates, and on those in the vicinity of the Indus and its tributaries in Sind and the Punjab, the presence of a generous water supply made possible the production of food on a large scale. The digging of irrigation ditches and the draining of marshes rendered necessary the pooling of the labor of adjoining communities and eventually the consolidation of social organization and the centralization of the economic system, with the accompanying centralizing of authority. Any men who had any ideas of cultivating their own plots were now coerced to work under threat of having their water supply cut off. Large surpluses of foodstuffs could now be accumulated and traded in exchange for foreign goods. Specialized craftsmen had to be trained to work the foreign imports. Soldiers were needed to protect the transport of exports and imports, scribes to keep records of the transactions, and state officials to carry on the business of the state.

All this had already been developed in Mesopotamia, in Egypt, and in the Indus valley by 3000 B.C. It was in the Fertile Crescent, in the alluvial plains between the Tigris and the Euphrates, that the great second revolution in the history of humanity took place, the Urban Revolution, the growth of the city and of the city-state. It was here that the urbanization and probably the dehumanization of man began, with the control of human beings as commodities, while at the same time the growth of the city made possible the development of all that we know as the "urbane" side of civilization. It is not the justness of a society's weights and measures which is the proper mark of its civilization, but its *kindness*. The overt expression of the inner civilization of a person is kindness; the same is true of the community.

TRANSPORTATION AND TRAVEL

If one moves from territory to territory, it is necessary to carry one's property with one. There are some nonliterate groups who possess so little real property that it can all be carried in one hand. The Australian aborigines, for example, make small net bags and a shallow wooden food container, which, apart from implements, serve to transport their worldly goods. Other hunting peoples, such as the Eskimos, have built the most elaborate and ingenious transportation devices, of which the dog-drawn sled is the most remarkable. The Eskimos also use their dogs as pack animals as the Plains Indians used to do. The Peruvian Indians used the llama. Indeed, every possible animal has been put to this use by man—from dogs to elephants. The dog-drawn travois was a device used by the Plains Indians. Two poles were harnessed to the animal with the ends dragging on the ground, and the load was put on the platform that rested on the poles. In Central Asia, the same device is used drawn by a horse or some other animal.

The wheel was never discovered in America. It is first known from a representation in Sumerian art from about 3500 B.C. Interestingly enough, the wheel was not used in Egypt till about 1650 B.C. The wheel revolutionized transportation and accelerated communications enormously.

Thus far we have talked about transportation on land. Marine and river transport are no doubt very old. Navigable river-boats are known from Egypt from 4000 B.C. and these were undoubtedly sailboats. By 3000 B.C. such boats were sailing the eastern Mediterranean. Water transport is at least as old as the Mesolithic and was already well developed by the Neolithic.

Hewn-out trunks of trees and rafts are used by the Australian aborigines, and the Eskimos use that remarkable invention of theirs, the kayak, which they manipulate with consummate skill.

The American Indian canoe is familiar to everyone. The outrigger canoe of Oceania is another remarkable invention of a nonliterate people. The outrigger was practically impossible to capsize, and it was possible to sail into the wind at an angle, an achievement unknown to Europeans before about the time of Columbus.

The most accomplished boatbuilders and sailors of the nonliterate world were those remarkable people, the Polynesians. Not only had they achieved great skill in boatbuilding, but their accomplishments were equally great in the science and art of navigation. This combination of qualities made them the most successful sailors the world has ever known, for they planned and achieved voyages by sea of two thousand miles and more. It is quite possible that on some of their voyages the Polynesians reached the coast of America.

EXCHANGE OF GOODS AND SERVICES

Barter, trade, and commerce largely depend upon a society's exchangeable surpluses. On the other hand, one can on occasion manufacture something that one knows to be valued by a neighboring group. Aborigines of northern Australia who possess a deposit of red ocher barter this with members of adjacent tribes for such things as weapons, fish, and yams. The ocher is carried to its destination either in lumps or in a fine, prepared powder. It is used for decorating the body, as well as implements and weapons, and is everywhere much valued. Indeed, every Australian tribe seems to have engaged in some form of barter, and such activities were often associated with certain ceremonies. For example, upon the death of a male, members of the tribe sent their spears to neighboring tribes who offered certain articles in exchange. These were then taken back by the original messengers and distributed among the owners of the spears. In this way, articles were obtained that were not manufactured within the tribe, and methods of making new implements were learned.

Barter is the direct exchange of goods for goods as such. Trade consists in the direct exchange of one value for another,

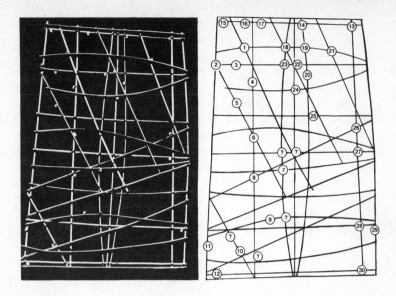

Micronesian navigation chart showing the Marshall Islands.
(Courtesy of the British Museum) In general the straight
horizontal and vertical sticks are primarily intended as supports
for the map, though incidentally they may represent the direction
of the swells. The diagonal and curved sticks represent the swells
produced by the prevailing winds which travel in a direction at
right angles to each stick on the concave side. The shells tied to
the sticks represent the position of most of the islands in the
group. The islands are remarkably accurately located.

either in terms of one another or some common denominator
such as money. Commerce consists in the interchange of commodities and embraces both barter and trade.

The form of barter known as silent trade is very ancient and
still widely practiced today. It consists in leaving a group of
articles in a certain place and then going away. Later the other
party arrives, examines the goods, and either leaves others in
exchange or takes his departure without disturbing the original
articles. The Congo Pygmies and their Bantu neighbors so
trade, as do the Reindeer Chukchee of Siberia with the maritime
Chukchee. This practice was also followed by Californian In-

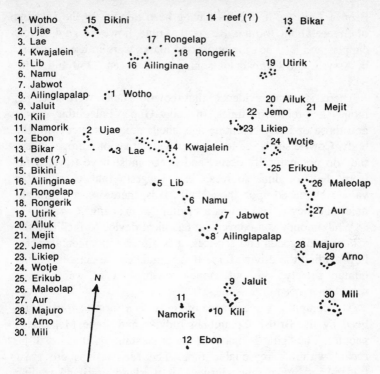

1. Wotho
2. Ujae
3. Lae
4. Kwajalein
5. Lib
6. Namu
7. Jabwot
8. Ailinglapalap
9. Jaluit
10. Kili
11. Namorik
12. Ebon
13. Bikar
14. reef (?)
15. Bikini
16. Ailinginae
17. Rongelap
18. Rongerik
19. Utirik
20. Ailuk
21. Mejit
22. Jemo
23. Likiep
24. Wotje
25. Erikub
26. Maleolap
27. Aur
28. Majuro
29. Arno
30. Mili

Chart of the Marshall Islands. The left-hand column refers to the
position of the islands as marked by the shells on the navigation
chart on the opposite page. (Courtesy of the Royal Anthropological
Institute, London)

dians, and in Malaysia, in New Guinea, and in other places.

Barter seems to be carried on more between tribes than
within the tribe. Face-to-face trading is the common practice
within nonliterate societies. In some societies some trade object
is used as a common denominator of value. This is known as
money-barter. In New Guinea and elsewhere cowrie shells may
thus serve, in the Philippines rice, salt in West Africa, tobacco
in Siberia, and metal objects in the Congo.

The use of money as such in nonliterate societies is limited. It
occurred in West Africa and the Congo, in Melanesia and
western North America. Coined money first appears in the Late

Bronze Age, the first coins having been minted by the Lydians of Greece about 700 B.C. Before that time, however, gold, silver, copper, and lead bearing impressed marks were in use as money in Assyria and Cappadocia some time between 2250 and 1200 B.C.

There is some evidence that cowrie shells were used as money in prehistoric times. In many Upper Paleolithic graves and those of Early Neolithic age, shells are associated with the body. Cowrie shells are found over wide areas of Europe, where they do not naturally occur, and hence must have traveled far from their original sources. This suggests that considerable value was placed upon these shells. It is, therefore, not improbable that they were used as a medium of exchange.

While money is a convenient technical device to facilitate the exchange of goods and services, it is also a fundamental factor in providing the driving force that stimulates the activities of all human society. Without money, civilization as we know it would have been impossible.

The compulsory acceptance of "gifts" is a form of trade practiced by the British Columbian Indians and some Melanesian peoples. The "gift" is made on the understanding that it will be repaid within a reasonable time. The Melanesians are great traders and make long journeys from island to island trading their goods in exchange for others. The profit motive is generally less involved than the prestige acquired in the process of trading.

Among the Zuñis of the American Southwest, the emphasis is upon cooperation in all activities. The people live by agriculture and stock-raising. Many have automobiles. In all transactions within the tribe, there is a complete absence of the profit motive. Objects have no fixed exchange value, "sales" do not exist, property is simply transferred within the framework of personal relationships. In a singularly inhospitable physical environment where land is scarce, land has no value based on scarcity because, according to the Zuñis, what *is* scarce and expensive is the labor to work the land. When a field is "sold," the size and location are of less significance than the kinship ties and relative ages of those engaged in the transaction. "I bought this field" (a good cornfield), said one Zuñi, "for a piece of calico because he was an old man," meaning that the old man had no further use for the field. It is not without significance that the women among the Zuñis play the dominant role in economic activities.

Markets develop out of the periodic gatherings of neighboring groups for the exchange or sale of goods. Middlemen in nonliterate societies are quite frequent. Thus, the Wishram Indians, fishermen on the Columbia River, collected goods from as far as California and the Plains. At definite times they would hold markets where Indians could buy fish, slaves, canoes, shells, horses, and so on. The Auki villagers on Malaita Island in Melanesia manufacture shell money, which they trade for food at the markets that are regularly held on the island.

Markets are widely held throughout Africa and may go on for days. In the southwestern Congo the Bakubas hold markets every three days, and by this regular recurrence of the market they measure their week. In the Uganda the chiefs supervised the markets, fixed the price of commodities, and claimed a 10 per cent tax on all sales.

In the Aztec capital, Tenochtitlán, where Mexico City now stands, local markets were held daily in various parts of the town. In addition to barter, articles such as beans, cotton cloth, copper ax-blades, and quills of gold dust served as media of exchange. These articles had certain standardized values which were regulated by special officials, who also supervised weights and measures and adjudicated disputes.

In addition to the immediate purposes achieved in the exchange of goods and services, trade has served to foster peaceful relations between peoples, to assist the spread of ideas and novel ways of doing things. Indeed, trade has been the great fertilizer, and the most complex civilizations have developed in those areas of the world which have been the great centers of commerce.

FORMS OF PROPERTY

The American historian James Harvey Robinson said, "The little word *my* is the most important one in all human affairs and properly to reckon with it is the beginning of wisdom." The ownership of property entails the exclusive use, enjoyment, and control of those things which are of value. Property is personal when controlled by the individual, common when a number of individuals have rights in it.

Among hunting peoples generally, while property rights exist, they are often much less exclusive than among ourselves. A wife and children are one's exclusive property, generally much loved and cherished, and not at all regarded with the callous indifference that has unjustly been attributed to men in nonliterate societies. Food is shared among the members of the group according to certain specific rules, the best parts usually going to the oldest persons. Food stored in caches may be used by anyone. One owns tools and implements as long as one uses them. A house belongs to the family that built it only as long as that family continues to live in it.

Among most hunting peoples land is held in common, and hunting and food-gathering rights belong to all. There are generally no private rights of this kind. This was also true among the pastoral Hottentots of South Africa, but if a man dug a well or opened a spring, it was regarded as coming under his special authority and, after inspection by the chief, was named after him. Anyone wishing to use the water had to obtain special permission from him, but it was equally his duty to see that no stranger or his stock was denied access to it.

The Hopi Indians, a horticultural people, own the land in common, title being vested in the clan, an enlarged family, and the land being distributed among the clan members for their use. As among the Eskimos, houses, tools, and other artifacts are owned by those who make them and use them. The right to property is its possession; there is no ownership which is not accompanied by use, and consequently no wealth which enables an individual to live off the labor of his fellows.

In nonliterate societies generally, the tribe is but an overgrown family rather than a state. Kinship ties and ceremonials hold it together, everyone is more or less related to everyone else, and the line of descent determines inheritance and social position. The compulsion of blood renders the terms "private" and "public" quite inappropriate. Common ownership is understandable under these conditions, with, of course, the usual rights of the individual to what he happens to have and is using. The mark of property becomes most manifest in intertribal relations. The tribe lays claim to its own hunting grounds, its real property, and also to forms of incorporeal property such as songs, dances, incantations, and gods. Incorporeal property rights have been, however, often vested in individuals who held the exclusive rights to certain chants or songs, myths, designs, poems, and the like. Usually no one ventured to indulge in any of these activities without special permission from the owner. An Andaman Islander, a Plains Indian, a Melanesian, and a Chukchee never ventured to sing the song of another without permission. There is often a divine sanction to the rights in a song or chant or dance, and this may be one reason why the right is so rigidly respected.

Those who know special incantations own a valuable property which, as among the Siberian Koryaks, they often sell at a good price. A vision that has been revealed to one, among the Plains Indians, for example, is so strictly one's own property that it cannot be transferred even to one's son without a formal sale. In matrilineal Melanesian tribes, however, it was obligatory to teach one's sister's son the magic spells.

Among the California Indians and some of the other simpler nonliterate societies, material property was not inherited, but destroyed at death. The Onas of Tierra del Fuego wrap a dead man in whatever clothes he may possess and burn him, his hut, and all his belongings. His dogs may be turned over to some kinsman.

With the accumulation of property that is highly valued, there is reluctance to destroy it upon the death of the owner, and rules are devised to regulate the inheritance of such property. The fact that spouses are members of different kin groups usually prevents their inheriting from each other. The individual in nonliterate societies rarely has the freedom to will his property to whomever he wishes—the social rules take care of that. In some societies inheritance passes through the female line, and this is known as *matrilineal*; when descent is reckoned in the male line it is known as *patrilineal*. This does not mean that in the first case males get nothing (or that in the second females are left nothing), but rather that the disposition of most of the property is in the hands of the females (or in those of the males).

We perceive, then, that the institution of private property is nowhere absent, but that it may be restricted to special things.

The more advanced the society is, the more exclusive property rights tend to be. The conception of property rights which prevails in any society reflects the social and political structure of the society and reacts upon those structures to create prestige, classes, and political power.

❧ 10 ❦

SEX AND MARRIAGE

PATTERNS

Every society makes certain fundamental distinctions that are based on the facts of nature, the two primary facts being sex and age. On the basis of sex, children are trained in particular ways, learning what others expect from them and what they may expect from others. This socialization process, as it is called, everywhere differs for the sexes.

Much, if not all, of what we regard as typically "masculine" or typically "feminine" behavior is culturally determined, so that what is esteemed to be typically "masculine" or "feminine" in our society may be quite the reverse of what is so regarded in another society.

The division of labor has already been discussed, but there are many other qualities which are inculcated in the sexes that are in some cultures considered feminine virtues and in others masculine virtues. Much of the differentiation between the sexes that was formerly taken to be biologically determined we now know to be culturally determined, determined not by nature but by nurture. In the Western world, particularly, we observe women daily performing tasks which only a few generations ago were generally considered beyond their capacities. With the in-creasing freedom granted to women, there has been a noticeable decline in the frequency of swooning and the "vapors," which seemed to be peculiar to the gentler sex in the Victorian period. There is scarcely an occupation that women in the civilized regions of the world do not fill, and fill efficiently.

In most nonliterate societies women are certainly regarded as the inferior sex and are less valued than men. The male by virtue of his greater muscular power is able to enforce his social superiority by brawn whenever it is challenged by the female, which is seldom indeed. Women have been forced to develop tactical approaches to the male of an indirect kind, to sharpen their wits, as it were, upon the whetstone of the male's orneriness. Hence, in all societies in which women are dealt with primitively (and this does not yet exclude our own society), women are alleged to possess greater cunning and shrewdness, to be more selfish, and to be more concerned with sex than the male. These traits are all culturally produced. Where women appear to be more concerned with sex than males, it is principally because the males expect the females to make themselves attractive. In fact, it is often necessary for females to compete with one another for the scarcer male's attention. It is evident that in almost all societies the male is more preoccupied with sex than the female.

SEXUAL BEHAVIOR

In almost all nonliterate societies premarital sexual freedom is more widely permissive than it is in literate societies. This freedom is considered to be an introductory preparation for marriage, and in many societies, as among the Muria tribe of India, such behavior is encouraged by the provision of appropriate boys' and girls' houses where each may take temporary partners.

In nonliterate societies there are generally no hard and fast rules concerning the number of wives a man may have; he may generally have as many as he can afford. Having more than one wife is an economic advantage, but it is often very difficult to obtain even one wife in such a society as that of the Australian aborigines, and there are few men who have more than one wife in Australian aboriginal societies. There are some societies, such as the Todas of southern India, in which one woman may have many husbands. This form of union is known as *polyandry,* the union in which a man may have more than one wife is known as *polygyny,* both states are covered by the term *polygamy,* and marriage of only one person to another is known as *monogamy.* Most nonliterate peoples live in monogamous unions.

MARRIAGE

Marriage is the social institution recognized by custom or law in which a man and a woman enter into a union with one another with the assumption of permanency. Everywhere the function of marriage is to provide a socially sanctioned, stable background for the mating of husband and wife so that they may produce and raise children. Indeed, marriage is the principal institution designed to ensure the continuation of the family and other groupings based on kinship.

In most nonliterate societies marriage is not a matter which is left to chance, as it is in most of our modern civilized societies, but is more or less strictly regulated within a definite framework. In all societies there is a proscription against marrying parents and brothers and sisters, and this is sometimes extended to cousins. There are, however, exceptions. For example, the kings of ancient Egypt had so hypertrophied a sense of their own importance that they felt that there was no one good enough, that is, of sufficiently pure "blood," to marry, so they married their own sisters. In fact, seven pharaohs, one after the other, did so, and it should be recorded that their offspring appear to have been of outstanding ability and intelligence. For the same reason such unions were favored by Peruvian and Hawaiian royalty. This suggests that the horror of incest is not inborn but represents an acquired distaste.

In many societies all marriages must be contracted with members of other tribes outside one's own group. Such marriage practices are known by the name of *exogamy*; marriage within one's own group is called *endogamy*. The origin of the incest prohibition has puzzled anthropologists for many years, and there have been many theories. One such widely accepted theory is that incest was an artificially created dislike designed to encourage marriages outside the group. It is assumed that the prohibition is very ancient, and that what it was designed to achieve was to place members of one's own group in that of a neighboring group to act, as it were, as ambassadors. Similarly, members of the other group performed the same function, and thus good relations were maintained between groups. It must be said that this is a conjecture. We shall probably never be able to ascertain the origin of the incest prohibition.

Another theory to which many authorities incline is based on the conjecture that early man almost certainly observed the effects that frequently followed upon close inbreeding within the family and between members of the extended family. Such close inbreeding often leads to abnormalities in the offspring, weaknesses of various sorts, and reproductive deficiencies. The prohibition of sexual relations within the family would therefore not only confer a selective advantage upon families and populations practicing the incest taboo, but would also serve to produce certain other benefits. It would solve the problem of sexual competition within the family. It would make outbreeding with members of unrelated families the rule and thus contribute to the cooperative binding of family to family, and of outgroup to outgroup. It would serve to train early the control of sexual expression. It would serve to provide the marriageable members of the family with mates very much more efficiently than could the family of which the individual was himself a member.

PREFERENTIAL MARRIAGE

In all societies the individual is responsible to his social group. In general the family constitutes the agent of the larger community, and it is through it that the behavior of the individual tends to be regulated. Marriage is one of the forms of behavior which tends to be very carefully regulated in all societies since marriage involves the creation of close ties not only between the marrying individuals, but also between their families. In marriage, kinship is established between two groups, not merely between two individuals. It may readily be understood, then, why each group should feel itself deeply involved in the marriage of any of its members. In our own society, parents and other relatives often see to it that marriages are contracted between their children and those of other parents who have had a similar background, or belong to the same "race," or ethnic group, or socioeconomic group, or religion. Most Catholics tend to marry members of their own religious faith; to that extent they practice a form of endogamy. Jews are similarly inclined to marry members of their own faith.

In many societies, cross-cousin marriage is obligatory. The term "cross-cousin" refers to cousins whose parents are brother and sister. If you are a male, then your cross-cousin is your mother's brother's daughter. If you are a female, your cross-cousin is your father's sister's son. Your father's brother, whether you are male or female, is your parallel uncle; your mother's sister is your parallel aunt. But your mother's brother is a cross-uncle and your father's sister is a cross-aunt. Parallel and cross-uncles and aunts are equally close relatives, yet a very large number of societies regard parallel kin as closer. Parallel uncles and aunts are frequently called "father" and "mother," and they in turn call their nephews and nieces "son" and "daughter," other terms being used for cross-relatives. The children of parallel uncles and aunts are often called "brother" and "sister" by their cousins, and since they are regarded as close "blood" relatives, marriage between them is considered incestuous and therefore forbidden.

In still other societies (the Miwoks of California and the Murngins of Australia), one may marry the daughter only of one's maternal uncle, but not the daughter of one's paternal aunt. In the Trobriand Islands precisely the reverse rule obtains. Such forms of marriage are known as asymmetrical cross-cousin marriages.

Parallel cousin marriage often takes the form of marriage of a male with his father's brother's daughter, as among the Bedouin Arabs of northern Arabia. This serves to keep the males within the band and thus to preserve a strong defensive and fighting force.

In many societies, upon marriage the wife is required to go and live in her husband's village. This is known as *patrilocal* or *virilocal* residence; the opposite condition, in which the married couple settles in the domicile of the wife's family, is known as *matrilocal* or *uxorilocal* residence.

The forms of marriage known as the *levirate* (Latin: *levir* = brother-in-law) and *sororate* (Latin: *soror* = sister) are very widely distributed. The levirate is the practice of marrying one's brother's—usually older brother's—widow. The sororate is the practice of marrying a younger sister of one's dead wife. The levirate was practiced among the ancient Jews, as the Old Testament tells us, and it is found also in many nonliterate peoples. Among the Chiricahua Apaches of North America both the levirate and sororate are practiced. Both the levirate and sororate may be obligatory, preferential, or permissive in different societies. The function of the levirate and the sororate is to maintain the bonds originally established between the two families, to ensure that they are not dissolved by death.

In many Indian tribes a man established a legal title, upon marriage, to his wife's sisters and kinswomen, and upon their attaining maturity he could consummate marriage with them—a custom known as *sororal polygyny*.

In some American Indian tribes, such as the Shoshone Indians of Idaho, and among the Australians and the Melanesians, *interfraternal* or *interfamilial exchange marriage* occurred. This involves a brother and sister of one family marrying a sister and brother of another family. "Take thou my sister and give me thy sister" was the formula used by the Arab people of Artas (Palestine).

MARRIAGE BY CAPTURE

In Europe the Yugoslavs, Orthodox as well as Roman Catholic, practiced forcible capture of girls for the purpose of marriage well into the nineteenth century. This often resulted in bloodshed. The Plains Indians frequently obtained wives in this manner, and until recently many such wives were still living among the remnants of our Plains Indians. Such wives enjoyed all the privileges of other wives.

Mock capture is practiced in many tribes, and this is, of course, by common agreement. It is found widely spread throughout Africa, and occurs in Melanesia and in China.

Where there is a surplus of females, groom capture may be practiced, as among the Kambots along Potters' River in New Guinea. The girl's father "kidnaps" a youth from a neighboring settlement. The victim is usually quite willing to be thus removed, and his father's show of wrath is appeased with gifts and a banquet. Since matrilocal residence, in which the groom goes to live in the locality of his wife's family, is not customary in this region, the plea of duress serves as an excuse for the breaking of the custom.

MARRIAGE BY SERVICE

In Genesis 29:18 Jacob says to his mother's brother, Laban: "I will serve thee seven years for Rachel thy younger daughter." But he was given the older daughter, Leah, instead, because the firstborn had to be given before the younger. So Jacob did a stretch of another seven years for Rachel.

Marriage by service is generally associated with matrilocal residence. This form of marriage is the regular practice among the Siberian Chukchee, Koryak, and Yukaghir tribes, among the North American Winnebago and Hidatsa Indians, and the loosely organized tribes of the Chaco.

In many tribes marriage by service is regarded as a substitute or alternative for the regular form of marriage.

BRIDE PURCHASE

In nonliterate societies the most usual way of obtaining a wife is by purchase. This does not mean that one buys a wife as one buys a mat or any other object. A purchaser of an object usually retains absolute rights to it. This is not so in the case of bride purchase or wife purchase. A man cannot intemperately abuse his wife without calling forth the wrath of his in-laws and, often, of his society. Husbands cannot sell their wives, even though they may otherwise dispose of their services.

While the payment for the bride may in part be regarded as a recompense to the family for the loss of her services, it is usually much more than that. The amount paid often determines the prestige that the bride will enjoy in her spouse's and in her own community. The purchase price endows the bride with a certain value. If the price is high, she thereby becomes a person not to be too lightly valued. The dowry that those of European descent provide for their daughters serves much the same function. In the latter case the husband gains the advantage, which he is expected to share with his wife; among nonliterate peoples it is the bride's family that gains the advantage. Marriage by bride purchase tends to be associated with patrilocal residence.

INHERITANCE OF WIVES

In some societies a son may inherit all his father's wives, with the exception of his own mother, upon the death of the father. This is known as *filial widow inheritance*. It was practiced among the medieval Mongols, as reported by Marco Polo, and also by the Araucanians of Chile. Among the Palvic and Bura tribes of Northern Nigeria, a grandson could inherit his grandfather's wives.

ELOPEMENT

Runaway matches have probably occurred in all societies, and in some it is the usual way of getting married. Among the Kurnais of southeastern Australia ten or a dozen couples eloped at a time. The older people tended to wink at these elopements, but, often enough, a father was very angry, even though he had himself married in the same way. With the passage of a little time, however, the eloped couple was readmitted into the tribe, perhaps first having to undergo a formal beating.

ADOPTIVE MARRIAGE

In Indonesia and Japan a man may obtain a wife by being adopted into the family. By this means, a patrilineal family (one that reckons descent through the father) without sons secures its continuance. By a legal fiction, the son-in-law becomes a "son." The fact that his wife thereby becomes his sister is conveniently overlooked.

Whatever form of marriage we consider, the function of marriage is everywhere the same: to cement the bonds of society through the strengthening of the ties between a male and a female who, in their reciprocal relationships, will combine both sexual and economic functions. In this way, through the family, which is the basis of society, the society will be maintained and controlled.

DIVORCE

Divorce is the dissolution of the marriage tie by competent authority. In most societies arrangements exist for the dissolution of marriage, and these are quite as various as the forms of marriage. In general, divorce is much less complicated among nonliterate than among literate peoples. There being neither religious nor governmental sanctions for marriage in the former societies, its dissolution is accomplished much more easily. The causes of divorce are many. In most nonliterate societies, adultery rarely constitutes cause for divorce, but where it is considered a cause, it is often punished. Barrenness is a frequent cause, and where a bride price has been paid, this is often returned to the husband or his family. Disease, laziness, and neglecting to provide food for the husband are widely encountered as causes for divorce. Among the Todas of southern India, there are only two reasons for divorce: one is that the wife is a fool, the other that she will not work. When a Toda man wants a divorce, he pays one buffalo to his wife's people, and in return he receives any buffalo he may have paid as the original bride price.

In general, divorce is no more difficult for women than it is for men in nonliterate societies, and in some societies women possess superior privileges in this respect. This is true among the Kwomas of New Guinea, the Dahomeans of West Africa, the Yurok Indians of California, and the Witoto of Brazil. Among the Arunta of Central Australia and the Baganda of East Africa, a man may divorce his wife at will, yet in neither society do women have the right of divorce. The Zuñi husband and wife have equal rights to divorce, but it is the woman who usually establishes it by merely depositing her husband's belongings outside the door. This was the case also among the Iroquois Indians of New York. Among the Chiricahua Apaches equal rights to divorce also obtain. It is interesting to note that in these three American Indian societies the status of women is quite high.

Divorce, like marriage, is a family affair, and in general families are inclined to discourage divorce, especially when the return of property is involved. On the whole, divorce is regarded as a regrettable necessity and a recognition of human frailty. In spite of the ease of divorce in nonliterate societies, it is by no means as frequent as might be supposed. Marriages are rarely entered into lightly, and the bonds established between husband and wife are not likely to be casually dissolved.

❧ 11 ❧

FAMILY PATTERNS

THERE IS EVERY REASON to believe that the family, in the form of a more or less permanent union between a male and a female and their children, is the oldest of human institutions. In spite of nineteenth-century theories, it is now generally accepted that there never was a period in human evolution which was characterized by promiscuous relations between the sexes. The oldest of all human institutions, the family promises to continue to endure as the basic institution upon which society rests, no matter what social changes may develop. For a time the Soviet Russians believed that they could eliminate the family, but they soon discovered that, at least for a whole nation, the idea was unworkable.

The *kibbutz,* or collective village settlement of Israel, formed originally as a communal society in which it was envisaged that both marriage and the family would be abolished, has served to do neither. What has been altered is the structure of the family, the children being brought up in communal centers by persons other than their parents. In the kibbutz the conjugal family is replaced by the extended family (see below). Whatever its faults the kibbutz seems to work very well, but whether it could ever serve as the model for a much larger society, a nation, cannot be determined until the experiment is tried. Meanwhile, however restructured and redefined, the family, in one form or another, continues to be the indispensable social unit.

The basic family unit is really constituted by the mother and child, no matter who the biological father may be. In fact there have always been such families, and they have generally managed to survive and do quite well. Sequoyah, the American Indian inventor of the Cherokee alphabet, was the product of such a family, and so have been many distinguished men and women of our own time. An adult male is not indispensably necessary for the survival of mother and child—but such a male can be, and usually is, of considerable help. Whether he be the biological father of his children or not, his role, as a member of the family through marriage, is essentially social. This has to some extent been recognized in the dictum that maternity is a matter of certainty, paternity a matter of inference. Patriarchal societies have understandably tended to play down the role of the mother and exaggerate that of the father.

The reciprocal advantages obtained by the partners in the marital union make the family, except in certain segments of highly sophisticated societies, a most appealing arrangement.

The family may, then, be defined as the social unit constituted by, yet distinguished from, marriage, the latter being the complex of customs which brings the family into existence.

There are three kinds of family: (1) *The conjugal family,* consisting of a married man and woman with their children, is the kind of family that exists among ourselves. It is sometimes called the *nuclear* or *biological family.* The conjugal family is always *bilateral,* which means that each spouse is, through marriage, a member of two families. The children similarly are bilaterally related to their parents' families. (2) *The polygamous family* consists of two or more conjugal families affiliated by several marriages, that is, by having a single married parent in common. For example, under polyandry, one woman plays the role of wife and mother in several conjugal families and by so doing unites them into a larger familial group. Under polygyny, the more usual type of polygamous family, one man assumes this role. (3) *The extended family* consists of uncles, aunts, cousins, parents, grandparents, and grandchildren. Unilaterality in family relationships, where a child is considered to belong to the extended family of its father, is termed *patrilineal;* on the maternal side, the unilaterality is recognized by the term *matrilineal.*

FUNCTIONS OF THE FAMILY

The principal functions of the family are the following: (1) sexual, (2) economic, (3) reproductive, and (4) educational. In the family union the powerful sexual drive is creatively gratified, reproduction of the group is accomplished, the conjugal group taken care of, and the potentialities of the children for participation in their community brought to fruition.

In every society the parents are held responsible for the care and upbringing of their children. In matrilineal societies, like those of the Trobriand Islanders and the Iroquois, the mother's brother will stand to the children as a father with respect to their socialization. Almost everywhere older children also participate in the socialization of their younger siblings.

THE SAMOAN JOINT FAMILY

In Samoa the unit of social and economic life is a household which consists of the members of a number of conjugal families joined together by patrilocal residence under the authority of one headman (*matai*). A single household may consist of as many as fifty persons, all related to the headman by "blood," adoption, or marriage. The headman is treated with great respect. If, however, he loses that respect, he is likely to be deposed.

Ten or more households unite to form a village with the kinship bond as their principal tie. The headman with the highest title in the village is recognized as supreme chief. Theoretically the joint family headman holds the power of life and death over the members of the household, but this he is seldom called upon to exercise. In everything pertaining to the household, he is the supreme authority.

Very early in his life the child is turned over to the care of his older brothers and sisters, who become entirely responsible for his care and discipline. By the time a child is six years of age, he has become an economic asset. At about their tenth year, brothers and sisters learn to become shy of one another, until they cease speaking to each other familiarly and avoid all contact with each other.

All the unskilled, petty, and humdrum tasks of life that can be delegated to children are so delegated by their elders, including the tending of babies. Every child is at the beck and call of all his relatives, who have the right to demand his services and to criticize and interfere in his affairs. By the age of eight or nine, boys manage to escape this grind by following their fathers and learning from them all that a boy should. Girls, however, cannot get away so easily. They can take out their first frustration in dancing, which they are encouraged to do, or they can escape from home for a time on the pretext of visiting relatives, with different groups of which they can live from time to time. In fact, rarely do children live continuously in one household. Children may be given away to friends or relatives by formal adoption.

THE POLAR ESKIMO FAMILY

Life for the Polar Eskimos is very precarious. Populations are small, and the family is small, consisting of husband and wife and their children, and perhaps an older relative or two. The latter will be supported only as long as they are capable of contributing to the livelihood of the family, for unless they are able to do so they become a serious liability that may threaten the survival of the whole. This the older people are aware of, and they often beg to be abandoned and left to die. Their relatives are usually reluctant to do this, though they may be forced to. For the same reason, if children come too quickly, they are often exposed to die. If her husband dies, a young mother may destroy her infant since the baby now has no means of support, and if the mother of a nursing baby dies, the baby is, for the same reason, buried with the mother, unless the father finds a woman in the community who can suckle the child.

This is not cruelty, but a regrettable necessity, for the Eskimos do not regard a baby as a complete human being until he is capable of sitting up. The Eskimos have the greatest love for their children. Mothers suckle their children for five years or even longer. No Eskimo child is ever corporeally punished— this is considered the greatest barbarism. Mothers instruct their daughters, and the fathers their sons, in the duties and privileges relating to their sex.

Practically everything that has been said of the Eskimos is true of the Australian aboriginal family. Among the Australian aborigines, much the same conditioning factors are operative to produce similar adjustments to an extremely inhospitable environment.

The conjugal structure of the Australian and Eskimo families, as well as that of the Bushmen of South Africa, is simple compared with the structure of the joint family of the Samoans or of the Chiricahua Apaches.

THE CHIRICAHUA APACHE FAMILY

Though the Chiricahua Apaches of the American Southwest are, like the Eskimos, the Australian aborigines, and the Bushmen, a hunting and food-gathering people, their land is a much more hospitable one than is that of any of the other three peoples. The populations of the Chiricahuas are therefore larger.

The household into which the child is born among the Chiricahuas is one of a cluster of families, each related unilaterally through the maternal line. Near an older woman and man live their unmarried sons and daughters, their married daughters and sons-in-law, their daughters' daughters (married and unmarried), and their daughters' unmarried sons. Upon marriage each daughter occupies a separate dwelling with her husband. Unmarried sons usually live in the parents' household, but unmarried adult sons may have their own adjoining dwelling.

The child's personality develops in relation to this large maternal extended family. Since the extended family from which the father has come may be located in the same vicinity, he may often see his paternal relatives, who invariably show a great deal of affection for him.

Discipline of the child is left to the parents, who take their duties very seriously. Children are rarely chastised, though on occasion they are whipped if all other means fail, but the point is made that adults do not like to do this.

The extended family, typified here by the Chiricahuas, is very frequent among nonliterate peoples. It is most likely to occur in regions where a band of families can live together, as among numerous African and American tribes.

THE CLAN

As a member of a family, one is bilaterally related to the families of both one's father and one's mother. Kinship is traced through both parents. There are some societies in which kinship is traced through only one parent, that is, unilaterally, either through the paternal or maternal line. Such a unilateral kinship group is known as a *clan* (sometimes called a *sib,* and when reckoned through the male line, a *gens*). A patrilineal clan is one in which all the members (at least in theory) are descended from a single ancestor through males. A matrilineal clan is one in which all the members are descended from a single ancestress through females. A clan which consists of the members of a single ancestor is a *lineage*. Some clans consist of a single lineage, while others, as among the Hopi Indians, consist of several lineages. Clans are always exogamous, and therefore husband and wife belong to different clans. One is born into a clan and, except in extraordinary circumstances, can never change it. One cannot voluntarily join a clan.

Families may die or dissolve, but the clan endures, in theory. In practice, in small populations, the clan frequently comes to an end because in a certain generation the children all belong to the sex that does not transmit the name—in which case they attach themselves to another clan and are adopted by it as real members. Certainly the clan appears to be more stable than the family in many nonliterate societies. If there are no siblings in one's family, one has siblings in one's clan. The clan serves as an auxiliary family, and no matter how remote clan members may be from one another in space, their kinship, as well as the duties and privileges that go with it, constitutes an indisoluble bond between them. Clan members are always expected to help each other, and in any issue they stand together against all other clans. Sometimes members of one clan are hostile to members of another clan, as among the Witoto of the Amazon, and in their own self-interest they may be opposed to the welfare of the group as a whole.

Clans do not appear until societies have advanced beyond the earliest stages of complexity, and they tend to disappear with the development of strong, centralized governments.

MOIETIES

When two intermarrying clans dwell together, each is called a *moiety*. Moieties may be exogamous, agamous (not regulating marriage), and endogamous. Exogamous moieties are common in Australia, less so in North America, and almost entirely wanting in Africa. The Canella and the Cayapo tribes of Brazil are characterized by the presence of agamous clans or moieties.

In societies in which there are many clans, the clans are

regarded as subdivisions of the moieties. Among the Seneca Indians, for example, there are two moieties, one consisting of the Bear, Wolf, Turtle, and Beaver clans and the other of the Deer, Snipe, Heron, and Hawk clans. Marriages were originally between members of the clans belonging to different moieties.

PHRATRIES

Where several clans consider themselves especially bound to each other and not as closely bound to other clans, they are, as a group, called a *phratry* (Greek: *phratēr* = brother). Subdivided moieties may be phratries, but a phratry need not be a moiety. It is simply a union between two or more clans. A tribe may have moieties and phratries.

The splitting of big groups into little groups, and these in turn into littler groups, is a tendency that operates in all human associations. But the reverse process is at least as strong, and it is undoubtedly the one that has been operative to the largest extent in the making of human societies. The family came first, and society developed out of the family.

THE FUNCTIONS OF THE CLAN

The functions of the clan are primarily: (1) to provide a bond of solidarity through the belief in their common descent, (2) to provide its members with all the necessary protections they might otherwise be unable to secure, (3) to regulate and control marriage, and (4) to adjudicate in disputes; and secondarily to provide the following sanctions for its members: (a) governmental, (b) economic, (c) religious and ceremonial, and (d) totemic.

The clan does not always function to the advantage of the society as a whole, and as society progresses, the functions of the clan are gradually taken over by the state.

METHODS OF RECKONING RELATIONSHIP —KINSHIP

When a child is born to a married couple there are at once established a number of special relationships to various persons on both the mother's and the father's side. In our society these relationships are restricted to the members of one's own family and the families of one's parents. While we emphasize "blood" relationships and relationships established by marriage, non-literate societies emphasize the *generation*. When we say "father," "mother," "son," and "daughter," as well as "brother" and "sister," we know that these are all relatives of the first degree; they are immediate "blood" relatives. In non-literate societies such immediate relationships are subordinated, and the notion of the generation to which one is related is elevated into what anthropologists call the *classificatory system*. In the classificatory system kinship is determined by affinity as well as by "consanguinity."

A member of the Arunta tribe of Central Australia, for example, uses the term "father" to designate his own father, his father's brother, his father's first cousin, and so on. All relatives in each generation are grouped into four categories. Everyone in the grandparental generation is either (1) *arunga*, father's father or his sister, (2) *apulla*, father's mother or her brother, (3) *chimmia*, mother's father or his sister, or (4) *ipmunna*, mother's mother or her brother. Grandparents call grand-children by the same terms that the latter call them.

As a matter of fact, the Australian aborigines in general do not distinguish between "blood" and marriage relationships. Among the Arunta and many other Australian tribes, relation-ships are primarily social rather than biological. For example, the mother's husband is considered the father of a child no matter who the actual genitor may have been. It is sociological paternity that is important, *not* biological paternity. The reason illegitimacy is condemned is not that it is immoral, but that the child has no legal father and therefore cannot be properly in-corporated into the society. However, among the Eskimos, chil-dren born out of wedlock are quite as enthusiastically received as children born in marriage. Such children generally remain with their mothers. If the mother is unhappy with the child, it will be adopted by someone else.

It would take us too far afield to consider other forms of classificatory relationship here. The important thing is that every relationship system in nonliterate societies plays a very real and vital role in regulating the conduct of individuals to-ward each other, in determining kinship behavior.

KINSHIP BEHAVIOR

In all societies the relations between kin are established by custom. Parents stand in certain relations to their children, and children must behave in certain ways toward their parents. In many societies, brothers and sisters after a certain age must reduce and modify their familiar relations to each other. In some societies, where matrilineal descent is the rule, the mother's brother assumes many of the duties of the father, as among the Melanesian Trobrianders and the Haida of the Northwest Pacific Coast. The specific relation existing between a mother's brother and his sister's child is known as the *avunculate*. In Melanesia the father's sister oversees the choice of a boy's mate, and he is expected to respect his maternal aunt more than his own mother. This relationship is known as the *amitate*.

In-law avoidance behavior is widely prevalent in nonliterate societies. Such societies appear to have solved the mother-in-law problem in the most absolute of ways. In a very large proportion of nonliterate societies, the son-in-law may not even look at his mother-in-law, and she may not even approach her son-in-law's hut. Among the Onas of Tierra del Fuego neither a man nor a woman may speak or even look at either of their parents-in-law. After a year the females may relax the prohibition on speech, but the prohibition remains absolute throughout the lives of the men.

Special privileged relationships in the form of *joking relatives* are found among many peoples. In different cultures the privilege of teasing each other, indulging in violent practical jokes and horseplay at each other's expense, is enjoyed by sisters-in-law, brothers-in-law, cross-cousins, paternal grandfather and grandchild, a potential husband and the woman he may inherit, and so on. American Indian and Australian aboriginal peoples excel at this sort of thing. The explanation for such behavior offered by anthropologists is that this privileged familiarity serves to compensate for the rather severe restraints which are put upon the behavior of most individuals within nonliterate societies as a consequence of the kinship system.

❧ 12 ❦

SOCIETAL CONTROLS AND
GOVERNMENT

In THE PRECEDING PAGES we have considered something of the significant groupings of individuals, that is, the social organization of nonliterate peoples. The social organization provides the framework upon which the individual can take his stand in relation to people. Every society has, in addition, developed means by which the behavior of the individual is controlled, that is, regulated, checked, and restrained. Under societal controls, mores and public opinion are the two great regulative forces; under government, political forms and judicial and legal procedures serve similar functions.

In all societies, social life, for the individual, represents an uneasy balance between his social obligations to others and the satisfaction of his own needs. Every society prescribes the forms which the indicated compromise shall take, in which the performance of social obligations is rendered more or less compatible with the satisfaction of personal needs. Every society permits a certain leeway on both counts, and deviation from the rules, within certain limits, is permitted by most societies.

Among the means of societal control are the statuses which go with the recognition of the stages or critical periods of life, as well as the more formal statuses and roles. These are not simply static recognitions, but dynamic roles, which provide a system of checks and balances which tend to keep the society in a feedback interactive equilibrium.

RITES OF PASSAGE

In every society the critical periods of life, such as birth, puberty, marriage, becoming a parent, becoming an adult, and death, are recognized as important events both in the life of the individual and of the community. These events are usually elaborately and solemnly celebrated with imposing ritual and ceremony. Such ritual ceremonies are known as *rites of passage*.

Three principal phases are recognized in all rites of passage: Separation, Transition, and Incorporation. For example, birth represents the separation from the womb of the mother, and the transition into that greater womb, the family and society into which the newborn is incorporated as a member of society. Marriage represents the separation from the single state through the transitional rites of sanctioned union to the incorporation into the married state with all its new roles and obligations. Puberty marks the separation from the phase of childhood into that of adolescence. Our confirmation ceremonies in the Western world constitute an example of a rite of passage.

Rites of passage are also performed in connection with the critical periods relating to the seasons. Since the changing seasons vitally affect the lives of everyone, it is of the greatest importance to ensure the proper passage of the one into the other, so that plants and animals will increase, and the rain will make the desert bloom.

Man is what he is because he is a highly imaginative symbol-making and symbol-using creature. It is by the imaginative use of these abilities that he deepens his understanding of the meaning of life, and enriches it, and by such celebrations gives dignity to the human condition.

STATUSES AND ROLES

Every person in a society is classified by social position. The social position occupied by a person is his *status*. The status of the person defines the *role* he is expected to play, the rights and privileges to which he is entitled, the duties he is expected to perform, and the manner in which he is required to conduct himself. Role is simply the dynamic aspect of status. The ability with which a role is performed is under the constant scrutiny and judgment of society.

Age, sex, kin, marital condition, leadership, class, caste, occupation, accomplishment, and the like, are some of the factors that determine status. Hence, an individual will have many different statuses. There are statuses to which an individual is born, such as sex, age, class, caste—such statuses are said to be *ascribed*. Accomplishment in an occupation, membership in an exclusive organization, confers *achieved status*.

The importance of status and role is recognized in every society, and hence its investiture ceremonies are designed to incorporate the individual into his new status and formalize the recognition on the part of all others of the privileges and obligations which attach to the prescribed status. As we have seen in our discussion of rites of passage, the individual changes ascribed statuses at various stages of his life, and he also changes achieved statuses, as when proficiency or uprightness wavers and the rust of time causes the hinges to creak.

Because status is highly valued in every society on account of the prestige it brings, there is a great incentive to achieve it. Hence, the members of a society will strive to excel in order to achieve a desired status and to maintain it. In this manner status becomes an important means of social control. Status is earned by the approbation or disapprobation of one's fellows.

Status and roles may be distinguished as (1) *basic,* as in sex and age, (2) *general,* as in occupation or accomplishment, and (3) *independent,* as in the leisure status of the clubman or society woman. In nonliterate societies basic roles linked to sex and age are important, while in more sophisticated societies such roles decline in importance and independent ones become more numerous and significant.

MORES

"Mores" (singular, "mos") may be understood to mean the moral standards of a group, the customs or folkways that are considered to contribute to group welfare. Every human group has found that there are certain patterns of adjustment to the environment (and the environment includes everything) which serve to make the wheels of society go round more smoothly, and these adjustments are incorporated into doctrines of group welfare. These doctrines are the mores of the group. By the standards which they set, they tell the individual, and the group, what behavior is right and what is wrong. The individual then knows what to do, and the society knows what to do when he doesn't do it. What may be considered immoral when the individual does it, as, for example, in the taking of human life, may be esteemed desirable when society sanctions it, as in capital punishment or in war.

In nonliterate societies the acts of the individual are believed by everyone to have consequences for the group as a whole; hence, the individual tends to regulate his conduct by the recognition of his social responsibility to the group. Punishment is likely to be immediate, not deferred to some afterlife. The individual is expected to be cooperative, altruistic, and noncompetitive; he is expected to be useful to his fellows, respect them, and contribute to their welfare. This view of human relations may or may not be extended to foreigners. In general, nonliterate people are very hospitable to men whom they have never seen before—after they have assured themselves that no harm is intended—but for tribes that are known to them, they may have great hostility and may even refuse to grant them the name of "men." Many tribes call themselves by names which mean in effect "we-are-men," implying that all others are not. For example, the term "Eskimo," employed for the first time in 1611 by the Jesuit R. P. Biard, was the name for these people bestowed upon them by the Algonquin Indians, and meant "eaters of raw meat." The Eskimos call themselves "Inuit," that is, "men," a name probably originally intended to distinguish themselves from animals. Many of the Eskimos who were first visited by white men expressed extreme astonishment to find that there were other men in the world, having been under the impression that they were the only human beings in existence! James Ross (1800–1862), the great Scottish explorer, was so greeted by one group of Eskimos who had never seen a white man before.

Similarly, the Hottentots call themselves "Khoi-khoin,"

meaning "men" or "proper human beings," implying that others are perhaps not quite so human. This ethnocentric viewpoint is not uncommon among technologically more advanced peoples. To the ancient Greeks other peoples were "barbarians," that is, "stammerers" who did not speak a sensible language. To the Russians the Germanic peoples were "mutes" ("nemtsy"), and to the French other peoples are "strange" ("étranges").

Among many nonliterate peoples it is the rule that territorial and hunting rights are respected by neighboring tribes. Anyone who breaks this rule is in some tribes considered subject to death upon sight. In such a case it is not merely the fact that a member of another tribe may be considered dangerous which is the cause of hostility toward him, but the fact that he has actually outraged a communal right.

All nonliterate societies expect children to pay attention when addressed, but in some societies this is a matter of morals, and the child learns very early, as in East Africa, that respect for his elders is paramount. If he is wanting in respect for his living ancestors, how can he be expected to respect those who are dead? Since the ancestors are very powerful the lesson is quickly learned.

Eskimos are among the most hospitable people in the world, and stinginess, greed, and selfishness are regarded as most reprehensible traits. When a man exhibits these to the extent of becoming a menace to the group, if he does not improve after having been given a chance to do so, he will be done away with. The man's kin will not seek redress, for this was an act committed in the interests of the group. But should an Eskimo kill another man in a fight, a feud between the families is precipitated, and atonement must be made.

As a rule nonliterate societies place great emphasis on sharing, and the selfish individual is generally condemned. Society is a cooperative venture, and the individual who does not cooperate is behaving contrary to the interests of all.

Kindness to animals is a fairly universal trait among nonliterates, especially pastoral peoples. When, not so many years ago, it was suggested to a Bantu chief that he yoke his oxen to the plow and so improve his agriculture, he replied, "How can I be so cruel as to make them work?"

In a certain sense kindness, indeed, is what morality is everywhere considered to be based on, and difficult as it may be for some to believe, kindness is what the law, at least ideally, also seeks to put into practice. But the practice generally falls short of the ideal, and it is the letter, especially in civilized societies, rather than the spirit of the law which tends to be enforced. Morality, in short, is the system of rules which gives significance to the activities of individuals in relation to each other. Moral behavior involves a consideration or choice of right and wrong, appraised in terms of a standard of values or a code of morals, toward which the individual recognizes a duty or feels a sense of responsibility.

PUBLIC OPINION

Public opinion is the expression of the moral judgment of a group. In nonliterate groups moral judgment in action constitutes public opinion. The exhibition of an attitude, the inflection of a word, or a gesture may suffice to exert the proper effect. A similar function is achieved in our own society by the great organs of public opinion such as journals, periodicals, and newspapers, and by newscasters and commentators.

The need for social appreciation is strong in all persons, being closely related to the basic need to be loved. All members of a society are greatly dependent upon their fellows for their own well-being, making public opinion a most powerful means of control. To lose face is equivalent in many societies to losing what is most precious, one's reputation, and in several societies, such as the Japanese, this calamity formerly obliged the victim to take his own life.

GOVERNMENT

Government means the authoritative administration of public affairs, an activity of persons whom the community authorizes to take action on its behalf. In all societies government is principally achieved through the agency of political arrangements and judicial and legal procedures.

POLITICAL FORMS

Politics, although another name for government, is generally recognized as that part of governmental affairs which has to do with civil administration. The earliest form of government arises within the family, the principal authority being the head of the household. This is very nearly the form of government which prevails among the Australian aborigines, the Eskimos, and the Bushmen of South Africa. These peoples are characterized by being small in numbers, without chiefs, and living nomadic or seminomadic lives. They live at a subsistence level and lack any form of organized warfare. Nevertheless, each of these subsistence societies has a council of wise men. These are the men who have shown wisdom in their conduct, and no matter how young or old they may be, they are the persons who convene and act as the government of the group when matters of weight are to be decided. The band, a group of associated families, provides the first occasion for government of the group. The band tends to be a politically independent organization.

With an established life in villages or settlements, sedentary populations assume a more politically interdependent organization. The factor of settled residence seems to be of cardinal importance in the increase in the complexity of political life. As one of the accompaniments of increased political organization, "graft" appears quite as often as it does in the so-called highly civilized societies. Government serves as a means of channeling collective action and social control, and this justifies it to the governed. At the same time those in authority find themselves in a position to do well by themselves. Power tends to exert a corrupting effect upon many human beings. In most societies as long as the rulers serve the community adequately, preserve law and order, and do not exploit their position too openly, the people are content enough to accept them for what they are. Government often comes to represent a struggle, a compromise, between the forces of good and those of evil.

In North and South America, government among nonliterates was mostly democratic. A good example is the League of the Iroquois, the federation known as "the Five Nations." The tribes forming the League were the Mohawk, Oneida, Onondaga, Cayuga, and Seneca. Each tribe maintained its independence in its own affairs, but in matters affecting the League, a council of fifty, drawn unequally from each tribe, acted for all. Public opinion was always given the fullest scope for expression before the council, whose decisions were arrived at by majority vote. Women named the members of the council, though no woman served on it. Women could also remove any member of the council at will.

Matriarchy (*mater* = mother, *archein* = to rule) in the commonly used sense of government by women, is a form of social organization which, so far as is known, has never existed, in spite of myths about the Amazons of Greece and anti-feminine remarks about American women. Matriliny, the transmission of names, inheritance, and descent through the female line, is not to be confused with the idea of government by females.

By means of governmental institutions, a number of communities may be united into a larger organized group. If the *tribe* is the social entity distinguished by its community sentiment due to a common culture, the *state* may be regarded as the political entity characterized by common governmental organization. State organization has been characteristic of many nonliterate societies, but these societies are usually of an advanced type. The Iroquois League provides an example of a nonliterate state, and so do the numerous highly organized populations of West, North, and East Africa. The forms of political control which states exhibit are termed as follows:

1. Oligarchy—supreme power vested in the hands of a small, exclusive class
2. Monarchy—supreme power vested in the person of a king
3. Gerontocracy—government by old men
4. Democracy—supreme power vested in the people and exercised by them directly or indirectly
5. Theocracy—government by supernatural direction, through priests or other sacred agents.

It must suffice to say here that every one of these forms of government is to be found among different existing nonliterate peoples.

JUDICIAL AND LEGAL PROCEDURE—THE LAW

Law is one of the several means of control, and there has probably never been an enduring society without some system of law. *The* law represents the measures taken by those in authority to enforce conduct deemed essential to the stability of society. *A* law is a rule of conduct imposed by authority, the breaking of which rule renders the individual liable to punishment. The rules and the punishments are accepted by the community as binding. Enforcement of the law implies that the community sanctions the application of physical force, whenever necessary, in fulfillment of the law.

In the simpler food-gathering societies, the force of public opinion, or a decision arrived at by democratic procedure, may be implemented by the authorized physical force of the group or some punitive measure. The community or the council acts as the court.

As societies progress, customary legal practices become codified into full-fledged laws and legal systems. Even in the most advanced societies, the court will usually recognize a long-established custom as having the force of law. In nonliterate societies no rigid distinctions are made—if any are made at all—between custom and law. Law may develop as a social necessity, but it is a necessity which the individual recognizes in all societies. In all societies the individual obeys the law not because he considers it right to do so, but because it is in his self-interest to do so. When it is not in his self-interest to obey the law, he will either break it or attempt to evade it, and if there are a sufficient number of individuals who feel strongly enough about it, the law will be changed. Thus the force of public opinion is stronger than any law, for it is public opinion that makes the law.

Law, like every other social institution, is functionally related to the structure of the society of which it is a part. We would consider it a crime to abandon a baby or our old parents, but the Eskimos consider this a perfectly proper and even kind thing to do. Murder is not a matter for group action among the Eskimos; it is privately avenged. Nevertheless, in spite of the fact that there are no rulers or headmen among the Eskimos, the members of an Eskimo group will combine to eliminate one of their own people who has become a nuisance, as we have already seen, or they will rise up as a group against evil-doing sorcerers.

A *crime* is a wrong committed against the community or state, a *tort* is a private offense, such as a trespass or nuisance, and though they may not give them such names, nonliterate peoples make the distinction, more or less. Thus, the Eskimo regards the uncooperative member of the group as a criminal. The murderer's offense is a private one until he commits a second murder, when the offense becomes a matter for the group. The evil sorcerer is a criminal; the felonious assaulter is not. A crime or a tort is whatever a particular society chooses to regard as such.

Not all societies consider witchcraft a crime. A Crow Indian simply utilized his own countermagic or that of a friendly shaman. Many tribes punish incest with death or expulsion, but the Crow Indians, though disapproving of the act, did no more than deprecate it. In England suicide until recently was regarded as a crime: self-murder. In the United States it is regarded as an occasion for sympathy. Civilized, as well as nonliterate, peoples differ among themselves as to what they regard as a crime.

In the simplest societies the judges are the people. Where the family is the law, it is the father who is the judge. Where a council of elders can be called into being, they constitute the judges, and such peoples as the Australian aborigines will hold trials by council and hand down judgments of which any civilized court might be proud. Trial by jury is quite recent, its beginnings going back to the tenth century in France, but trial by court is found among many nonliterate peoples. The council of the Australian aborigines acts as a court, and so does the tribal council of an American Indian pueblo. The court, in all probability, is of very great antiquity, for wherever several duly empowered individuals get together to hear a case and deliver a judgment, there is a court.

Among the Yurok Indians of California an aggrieved Yurok would engage the services of two nonrelatives from another community, and the defendant would do the same. These legal representatives were called "crossers," because they crossed back and forth between plaintiff and defendant. The two parties involved, the litigants, never saw each other during this time. After hearing all the testimony from each side and pleading the relevant law, the crossers rendered a decision according to the rule in such cases. For their efforts these "lawyers" received a piece of shell currency called a "moccasin."

Regular court sittings are not held among most nonliterate peoples, but they are typical of many African societies. The sittings often afford opportunities for the exhibition of forensic eloquence and also frequently serve as general entertainment for the public. In the simpler African societies, the public attending the legal sessions may ask the litigants and the judges questions, but the judges decide upon and render the verdict. In the more complex African societies such as the Ashanti (numbering over 200,000 members), judicial procedures were highly developed. Among the crimes that the Ashantis punished by death were murder, attempted suicide, certain sexual offenses, certain kinds of assault, certain kinds of stealing, cursing the chief, treason, cowardice, sorcery, violation of royal legislation, violation of a tabu, certain forms of abuse, and a woman calling a man a fool!

The king had the power of life and death, and no one might exercise that right but he. Private disputes could be settled between the household heads of the disputants. The tribal chief usually presided as high judge of his particular village or of the group of villages under his jurisdiction. As high judge he was assisted by associate judges consisting of the elders of the group. But the supreme court justice was the king. Since the Ashanti kingdom was a constitutional monarchy, the king, as in most monarchies, had the power to pardon and to modify a sentence, which, for a consideration, he not infrequently did.

Among the Jagga of East Africa it is considered unseemly to bring a private issue before the court. Public opinion holds that proper decorum requires that the plaintiff summon the defendant to the public common, where the community, presided over by the headman of the village, listens to both sides and then delivers an umpire's judgment. Payment of fees is eliminated, though a gift to the headman and a beer feast to one's neighbors are customary. There is no physical enforcement of the decision. The tribal chief takes no part in the proceedings, but is usually informed of what has occurred. A regular trial before the chief's court is an elaborate affair, and verdicts are strictly enforced.

Everywhere the function of the law is to maintain order within society so that individuals may know what their rights, privileges, and duties are; this implies, of course, the recognition of the rights, privileges, and duties of others.

The functions of law, then, are: (1) the ordering of relationships between individuals, and between individuals and social

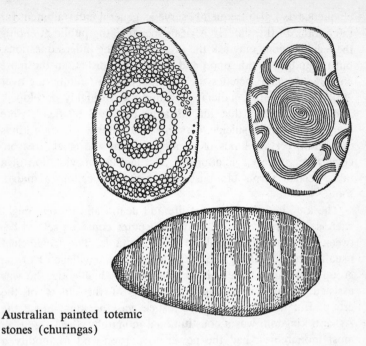

Australian painted totemic
stones (churingas)

groups, (2) the enforcement of the orders agreed upon by the community, (3) the just settlement of disputes between individuals and of offenses against the community, and (4) the adjustment of the rules of law to the changing conditions of life; the latter is often achieved not by changing the law but by changing its interpretation.

THE NATURE OF RELIGION

The individual in society at once feels very close to and very far from other human beings, but always there remains the strongest of desires, to be related to one's fellow man. Human beings have devised no more successful means of achieving this relatedness than religion. When one combines the emotional experience of the world as a mystery with the craving to solve some part of that mystery by identification with the powers that be, and with the feelings of relatedness and loneliness, one understands something of the matrix out of which the religions of all peoples grow.

The Australian aborigines are a deeply religious people, and like some others, they have no gods. If you ask the Central Australian aborigines of the Arunta tribe, Who created the world? they will tell you that in the *alchera*—that is, in the far, far away "dream time" (meaning long before human memory) —there dwelt in the western sky two beings, who are described as *Numbakulla*—that is, self-existing beings who came out of nothing. *Numbakulla* are formless beings. They did not create the world—the Australian aborigines do not presume to know who did; it all happened so long ago, in the time of the dreaming, in the *alchera*. It happened one day that the *Numbakulla* discerned, far away to the east, a number of *Inapertwa*, rudimentary human beings or incomplete ones, who possessed neither limbs nor senses, who did not eat, and who were all doubled into a rounded mass, in which just the vague outlines of the various parts of the body could be seen. These *Inapertwa*, who were destined to be transformed into men and women by the *Numbakulla*, represented the intermediate stage in the trans-formation of animals and plants into men. When the *Numbakulla* came down to earth and fashioned the *Inapertwa* into human beings, each individual naturally retained an intimate relation-ship with the animal, plant, or natural phenomenon of which he was indirectly a transformation, and with which he was at one time identical. It was in this way that human beings came into existence, and it is for this reason that each aboriginal possesses a *totem*, that is, an aimal, plant, or natural phenomenon—such as water, wind, sun, fire, cloud—with which the individual is closely identified. It is to that plant, animal, or phenomenon that the aboriginal believes himself to owe his original creation.

Numbakulla no longer exist, but the aborigines still celebrate them in religious ritual, though they do not offer any sacrifices to them or worship these no longer existing beings. Even though the aborigines have no gods, they have a full-fledged religion in their cosmogony, in their belief in guardian spirits, in rite of prayer, and in their belief in the existence of the soul. The criterion of the sacred for the aboriginal is the totem, but it is not the totem as such that is sacred, but what might be called its principle or spiritual essence. It is this essence which renders the totem sacred. Emile Durkheim claimed that totemism is the most primitive form of religion, yet we know of many people who exhibit no evidence of ever having passed through a totemic stage of development.

SPIRITUAL ESSENCE—MANA

The belief in a spiritual or totemic principle that we encounter in Australian aboriginal religious belief, we also encounter among many other nonliterate peoples. The belief in such a supernatural power was first described by Bishop R. H. Codrington in 1891 as occurring among the Pacific Melanesians. These people called that supernatural essence *mana*. Mana is a supernatural power or influence which is responsible for everything beyond the power of men, and it is something that works quite beyond the common processes of nature. "It is present in the atmosphere of life, attaches itself to persons and things, and is manifested by results which can only be ascribed to its operation. When one has got it he can use it and direct it, but its force," wrote Codrington, "may break forth at some new point; the presence of it is ascertained by proof." For example, a stone in the rough shape of some animal, tucked under one's belt, will bring success in the hunt, because of its wonderful power, the mana. A stone with the shape of a fruit, buried in the ground, will greatly increase the normal yield. Anyone and anything can have mana. There is no predicting what will and what won't have mana, but one thing is held to be certain: some things and some people, for some unknown reason, just don't have it. It is a matter of luck, but it certainly is not luck itself. The idea of luck largely does not exist among most nonliterate peoples. The element of chance as it is understood among ourselves is largely unknown, for in nonliterate societies everything, as a rule, has a determinate cause. Mana is an attribute or quality of supernatural power of the thing itself, or becomes so by attaching itself to anything. And it works of itself; one does not have to manipulate it or enter into any relation of any kind with it. It is this difference which distinguishes it from *animism*, the belief in spirits with whom one has to enter into relation, and from *animatism*, which simply assumes that everything is alive. Mana is powerful of itself; spirits are powerful only in certain relations.

Similar beliefs are found all over the world, though they may be called by very different names. Mana, it should be clear, does not imply the existence of supernatural beings but refers to an impersonal, wonderful power, which may be likened to a supernatural voltage with which the universe is imagined to be charged. Many nonliterate peoples do not make a clear distinction between this power as personal and the same power as impersonal. Among Indian tribes, the terms *manitou* (Algonquian), *orenda* (Iroquois), and *wakanda* (Sioux), which are different names for a power similar to mana, may be either personal or impersonal and may mean either supernatural power in the abstract or a specific supernatural being. Sometimes these terms may be applied to a holy man or a religious practitioner. In one form or another, the idea of a wonderful supernatural power is found among most peoples. Furthermore, man's attempts to relate himself to this supernatural power, and the beliefs that grow out of these attempts, lead to religious behavior.

THE TECHNIQUES OF RELIGION

In order to make the world intelligible to himself, man creates it anew in terms of his own experience; he brings the world, as it were, under domestication. The figures he projects into it are images compounded of the moral necessities which the authority figures of his experience, whether in the form of parents, teachers, rulers, or internalized as conscience, have urged upon him. As Ruth Benedict (1887–1948) has put it:

He sees in the external world the playing out of a human drama actuated by moral significance; that is, he sees it humanly directed toward rewarding those who have performed their required obligations and denying those who have failed in them. He is no longer in a blind and mechanistic universe. This wishful thinking, which is embodied in the religions of the world, and is worked out conceptually in mythology and theology and behavioristically in the religious techniques of petition and rapport, ranks with the great creations of the human mind.

Inherent in these religious techniques is the imperative urge of the believer to act—to act according to the requirements of his beliefs and according to the nature and meaning of the universe as revealed to him. The urge to act in this way constitutes the moral and social implications of all true religious experience, and gives rise to the religious techniques that are almost universally found in human groups: *magic,* the attempt to compel the supernatural; *reverence,* the combination of awe with love and admiration which one offers to the supernatural powers; *divination,* control by foreknowledge of supernatural power; *sacrifice,* control by gift, in which one puts the supernatural powers under compulsion to repay the gift by doing what one wants, or else simply desires to please the supernatural powers; *tabu,* control by abstention because one fears the supernatural powers; *fetishes* and *amulets,* two aspects of religious behavior—supernatural power treated as an attribute of objects makes amulets of those objects, and when the objects are treated as the seat or house of spirits, they constitute fetishes; *prayer,* the communication with the supernatural powers through speech. In addition, there is the belief in *guardian spirits,* widespread among the American Indians, where one may either place oneself in communion with a guardian spirit and at his service, or he places himself in communion with you and at your service.

Compulsion and rapport are the two extremes of religious behavior by means of which one enters into relation with the supernatural powers. In between these extremes there is every possible degree of compulsion and rapport and combinations of both. The techniques mentioned are the forms of behavior which are to be found in all the religious systems of which we have any knowledge.

CEREMONIALISM AND RITUAL

The forms which religions assume in different groups are as various as they could possibly be, and so is the pervasiveness of religion in any culture. Among the Polynesian Samoans supernaturalism is at a minimum, whereas it virtually occupies the whole life of the Polynesian Maoris of New Zealand. On the whole, nonliterate peoples are more profoundly religious than civilized ones. Among the Zuñis of the Southwest, religion is highly formalized, whereas the practices among the Chukchee of Siberia are relatively informal. This brings us to the matter of *ceremonialism.* Ceremonialism is not inevitably rooted in religious faith, but it is a form of behavior which is usually, though not always, associated with religious acts. It may be defined as a body of formal, sanctioned observances, learned by observation or precept, that indicate an attitude of reverence toward the supernatural. Ceremonialism takes on the character of *ritual* when the order of words or the pattern of behavior is thought to have inherent virtue or power to produce results. In its secular form, ceremonialism becomes pageantry. Secular pageantry has, of course, often been taken over for the purposes of religious ceremonialism. The function of both ceremonialism and ritual is essentially to provide a collective or group participation in the solemnization of collective behavior, to yield a collective sense of the importance of that behavior.

RELIGION AND MORALITY

While the religious systems of the Western world, indeed, all the higher ethical religions, are concerned with the conflict between good and evil, many of the nonliterate religions are more or less unconcerned with this aspect of life. In such cultures the relatedness which the individual feels toward the supernatural powers may, as far as his fellow men are concerned, be dealt with in secular terms. And ethical sanctions may be largely a matter of secular concern, though religious sanctions may often be massed behind them at many points. Religion and morality are not, however, identified in such cultures. At the present day, we are witnessing in the Western world, at one and the same time, the emergence of a strong secular conception of morality based on science and experience and a return to revealed religion and some of its more secularized counterparts, such as the various Reformed versions of Christianity and Judaism.

THE MEANING OF RELIGION

Everywhere men have evolved religious systems in which religious behavior has repeatedly and independently been calculated to secure similar ends. This fact bears testimony to the unity of the human mind. Everywhere the religious experience creates the atmosphere and attitudes enabling human beings to regulate their conduct in the world in which they find themselves. With the development and deepening of the meaning of the religious experience within the matrix of an increasingly complex social world, human beings begin to understand that the communion with the supernatural powers must be extended to the communion with one's fellow human beings, and finally to a moral obligation of fellowship that is universal.

The anthropologist observes that the freedom to develop and subscribe to any religion one chooses in a democracy leads to the democratization of religion, and thus gradually to closer understanding among people, which unbending orthodoxies have in earlier times successfully prevented. In the religions of the world the anthropologist sees an expression of the unceasing struggle of humanity toward the attainment of the community of man, the reverence for life, and the destiny of man which, in a mysterious universe, may gradually lead men to the discovery that the way of humanity must inevitably be through the path of cooperation.

≫ 13 ≪

MYTHOLOGY AND

PHILOSOPHY

W HAT IS MAN? What is life? What is death? These are questions that every people has asked and attempted to answer. When the question relates to the supernatural, to the great questions of creation, the answers usually take the form of a myth. A myth may be defined as a traditional story, accepted as historical, embodying the beliefs of a people concerning the creation, gods, the universe, life, and death. Legends and folk tales are often hard to distinguish from myths. When stories account for the origins of things and are believed to have taken place in a remote (mythical) past, they are classed as myths. Legends may or may not be based on historical events, and whatever they describe generally refers to historic time. Folk tales, or the folklore, of a people are a people's unrecorded traditions as they appear in its customs, beliefs, magic, and ritual.

THE FUNCTION OF MYTH

The worldwide study of myths makes it abundantly clear that man is a myth-making animal. His myths indicate the play of the imagination upon the raw materials of life, and the conversion of these, by a wishful-thinking imagination, into acceptable explanations and reassurances. In a profound sense the imaginative life of a culture is its most living reality. In myth, the universe and everything in it are remodeled to the heart's desire. The myth-makers usually elect themselves as the chosen people or as the first and originally created people, and they recast the universe to function, as it were, in more humanly controllable terms. The discomforts of life and the mysteries of the universe are, in myth, rendered understandable and even justifiable; hence, life is rendered more secure and tolerable than it would otherwise be.

Just as we sometimes tend to daydream, imagining ourselves in all sorts of pleasant situations, so the tendency of the human mind has been to institutionalize the "group daydreams." This tendency exhibits man, to some extent, in control of the universe by his habit of explaining it.

Mythology and folklore are not philosophic attempts to explain "the nature of things," though the desire to explain is usually there, but they are to be understood rather as arising out of man's desire to bring the universe into some sort of manageable configuration through the play of the imagination.

THE FUNCTION AND DISSEMINATION OF FOLK TALES
The function of folk tales is very similar to that of myths. In the folk tale we are dealing with a novelistic activity among nonliterate peoples, which occurs, of course, without benefit of the written word. The free play of the imagination upon the everyday experiences of life is sufficient to account for the origin of folk tales.

Folk tales have been studied by experts the world over, and it has been found that many elements in tales, and also complete tales, have traveled over great distances and been incorporated into the cultures of many different peoples. Some tales have even been traced to their origin, and then followed throughout the world. For example, in *Grimm's Fairy Tales* there is the famous story of "The Musicians of Bremen." This is the story of the despised and rejected animals who set out for Bremen where they propose to become State Musicians. They frighten a group of robbers out of their hut (the robbers symbolizing mankind), and in spite of the latter's attempt to regain possession, they live on in the hut happily ever after. This tale has been traced from its origin in Central Asia, through Europe, India and the Indies, China, Russia, and among the American Indians. Parts of the tale undergo change, but in essentials it is generally recognizable as the same one.

Every culture adapts its tales to its own peculiar needs, and the same is true of myths. Hence, the study of folklore is most helpful in tracing the movements and contacts of peoples who have no written or other records of such movements or contacts. Folklore is one of the means by which the anthrolopogist is able to trace the *diffusion* of cultural traits.

MYTHS, FOLK TALES, AND CULTURE

Since it is found that myths and folk tales are constructed out of the materials of everyday life, it is obvious that these forms of artistic activity are extremely valuable in throwing light upon the dominant cultural interests of a people. In the mythology and folk tales of a people we are invited to view both the everyday events of the life and the spirit in which those events are combined and lived out. An expert, by interpreting myths and folk tales, can often give a fairly good account of a culture, in some of its aspects at least, without knowing anything of the source of the myths and folk tales. The Pueblo Indians, for example, place much emphasis upon space relations in their daily lives, and this is reflected in their mythology. The concern of the Australian aborigines with animal life informs every aspect of their mythology.

When the language and imaginative genius of a narrator and the colloquial rhythms of his style and energy are captured in print, the imagination and spirit of the culture is brought to life.

PHILOSOPHY

Nonliterate man casts the net of thought over the whole world. Mythology and religion may be closely related, but where one grows out of man's everyday life, the other grows out of his concern with the supernatural. And so it is with his view of the world, which will be compounded of secular, religious, mythological, magical, and experiential elements all rolled into one.

Most nonliterate peoples are extreme realists. They are bent on bringing the world under control; and many of their practices are devised to insure that reality will perform according to their bidding. In the conviction that the spirits are on his side, a man may then make all the necessary preparations for the success of an expedition. Coercing reality to do one's bidding by manipulating it in the prescribed manner is, for the nonliterate, a part of reality.

It is necessary to understand that nonliterate peoples identify themselves very much more closely with the world in which they live than do the literate peoples of the world. The more "literate" people become, the more they tend to become detached from the world in which they live and alienated and disengaged from the world of nature.

What happens *is* reality to the nonliterate. If ceremonies calculated to increase the birth of animals and the yield of plants are followed by such increases, then the ceremonies are not only connected with them but are part of them; for without the ceremonies the increase of animals and plants would not have occurred—so the nonliterate reasons. It is not that the nonliterate is characterized by an illogical mind; his mind is perfectly logical, and he uses it very well indeed. An educated white man finding himself suddenly deposited in the Central Australian desert would be unlikely to last very long; yet the Australian aboriginal manages very well. The aborigines of all lands have made adjustments to their environments which indicate beyond any doubt that their intelligence is of high order. The trouble with the nonliterate is not that he isn't logical, but that he applies logic too often, many times on the basis of insufficient premises. He generally assumes that events which are associated together are causally connected. But this is a fallacy which the majority of civilized people commit most of the time, and it has been known to happen among trained scientists! Nonliterates tend to adhere too rigidly to the rule of association as causation, but most of the time it works, and by the pragmatic rule what works is taken to be true.

Nothing could be further from the truth than the idea that nonliterates are utterly credulous, superstition- and fear-ridden creatures, without any capacity or opportunity for independent and original thought. In addition to good horse sense, the nonliterate usually displays much practical sense based on an appreciation of the hard realities of life. Listen to this American Indian, an intensely religious individual, recounting the essence of the instruction current in his tribe:

Help yourself as you travel along the path of life. The earth has many narrow passages scattered over it. Some day you will be journeying on a road filled with obstacles. If then you possess the means for strengthening yourself, you will be able to pass through these passages safely. Indeed, if you act properly (i.e. circumspectly) in life, you will never be caught off guard.

As Paul Radin, to whom this was told, remarks, scarcely anything more practical than this could be imagined.

Nonliterates are obliged to be practical; their very lives depend upon being so. In order to make such practicality acceptable to the younger members of the group it has been necessary to endow it with supernatural sanctions and to surround it with myth and legend. There are good, sensible reasons for many of the nonliterate practices that more sophisticated peoples tend to regard as either stupid or barbaric. Such practices are generally neither stupid nor barbaric but serve some very definite purpose.

The individual in nonliterate society is generally a much more closely knit member of the group then his opposite number in civilized societies of the Western world. But, contrary to general belief, the nonliterate usually enjoys a considerable amount of freedom of personal expression. There are many ways in which the individual can so express himself and at the same time contribute to the welfare of the group. The form of self-expression with which we are concerned at the moment is the philosophical variety.

In every nonliterate society we find at least as many philosophers as we do in the civilized societies of the Western world. Everyone is, of course, to some extent a philosopher, but the general opinion that such persons are rare in nonliterate societies requires particular correction. Indeed, in many a nonliterate society there are one or more doubters, usually in the older age grades, who are rather skeptical about some of the traditionally accepted beliefs. In *Primitive Man as Philosopher,* Paul Radin gives this commentary of an Amazulu tribesman of East Africa on Unkulunkulu, the supreme deity of the people:

When black men say Unkulunkulu or Uthlana or the Creator, they mean one and the same thing. But what they say has no point; it is altogether blunt. For there is not one among black men, not even

the chiefs themselves, who can so interpret such accounts as those
about Unkulunkulu as to bring about the truth, that others too may
understand what the truth of the matter really is. But our knowledge does not urge us to search out the roots of it; we do not try
to see them; if anyone thinks ever so little, he soon gives it up and
passes on to what he sees with his eyes, and he does not understand
the real state of what he sees. Such then is the real fact as regards
what we know about Unkulunkulu, of which we speak. We say
we know what we see with our eyes, but if there are any who see
with their hearts they can at once make manifest our ignorance of
that which we say we see with our eyes and understand too.

As to our primitive condition and what was done by Unkulunkulu
we cannot connect them with the course of life on which we entered when he ceased to be. The path of Unkulunkulu through our
wandering has not, as it were, come down to us; it goes yonder
whither we know not.

But for my part I should say, if there be anyone who says he can
understand the matters about Unkulunkulu, that he knows them
just as we know him, to wit, that he gave us all things. But so far
as we see, there is no connection between his gift and the things
we now possess.

I say then that there is not one amongst us who can say that he
knows all about Unkulunkulu. For we say, "Truly we know nothing
but his name; but we no longer see his path which he made for us
to walk in." *All that remains is mere thought about the things we
like.* It is difficult to separate ourselves from these things and we
make him a liar. For that evil which we like of our own accord,
we adhere to with the utmost tenacity. If anyone says, "It is not
proper for you to do that; if you do it you will disgrace yourself,"
yet we do it saying, "Since it was made by Unkulunkulu where is
the evil of it?"

Many other examples of acute criticism could be cited from
nonliterate peoples but space forbids. It should, however, be
added that in nonliterate societies professional critics as such do
not exist. The professional critics come into being for the first
time when life becomes easier and men can find opportunities
which will enable them to live and to be critical.

The right to freedom of opinion is strongly maintained in
nonliterate societies as a rule. As among ourselves, the opinions
held must not be expressed in bad taste nor must they outrage
the received canons of decorum. A man's thoughts are real
events, and they must be respected, however strange they may

appear. Thought is, to the nonliterate, as much a part of reality as is anything else that is real. As long as the individual's reality does not interfere with anyone else, he has the right to express himself freely. A man's personality must not be affronted or unnecessarily curtailed.

Since thought gives expression to the reality of life beneath the surface, it is considered as important to permit its adequate functioning as it is to encourage it. The right to freedom of thought is extended also to children, and one of the most striking facts reported by anthropologists is the respect that is almost everywhere offered children as personalities in their own right in nonliterate societies. Radin recounts how on one occasion, among the Winnebago Indians, desiring to purchase a pair of child's moccasins, he approached the father of the child on the matter. He was told that the moccasins were, of course, the child's. Upon being pressed, the father agreed to consult the child, who was about five years old, as to whether he cared to part with them. "The whole transaction took place in a perfectly serious manner. There was not the slightest flippancy about it. The child refused and that ended the matter."

It is the same respect for children which, doubtless, causes the Winnebago to assert, "If you have a child, do not strike it. If you hit a child, you will merely put more naughtiness into it."

Most nonliterate people take the view that the individual is responsible for his own conduct. His parents can provide him with the idea and the model of what that conduct should be, but it is up to him to do something about it. A supreme obligation of parents, therefore, is held to be their responsibility as educators of their children in the ideals and requirements of a good member of the group. Such a good member must not only be a cooperative person, but the male must also be proficient in the ways of earning a living, tracking, hunting, and other manlike activities, while the female must likewise be accomplished in keeping house, food-gathering, and other such feminine activities. Parents who tend to fall short of their duties are met with much public criticism, and public opinion is usually a force sufficient to cause them to mend their ways.

Strength of character, wisdom, and loyalty to one's group are the principal traits emphasized in nonliterate societies. And then there is the sense of proportion which seems to be so marked a

characteristic of the nonliterate's world view—the proper order
of things in relation to each other. One must learn the limits of
the natural order, not only that one may control it, but that one
may obey it.

There is one thing that is somewhat difficult to control, and
that is death; but one can obey it. One can also make attempts
to control it, and most peoples, nonliterate and literate, have
attempted to do so through the doctrine of immortality, the
belief in a life after death. Immortality, with its promise of a
usually happier afterlife, has served to ease the sojourn on earth
of many a human being, but, on the whole, nonliterates have a
very realistic view of death. The concept of immortality is obviously born of wishful thinking, for death is a very evident and
unavoidable reality. The problem, therefore, is to find some
means of rendering it acceptable, of perhaps denying its reality
altogether.

The Aivilik Eskimos, for example, do not regard death as a
hard, unbearable fact. It is like sleep; after a little time, the
body reawakens. When the body ceases to breathe, they regard
this as but an event in a never-ending cycle. Death, like birth,
is only an event in time, and life is above time. They maintain,
therefore, that they can run all risks, squander their possessions,
as well as their lives, because they are immortal; for life extends
beyond death, beyond the corruption of the body. Indeed, the
most difficult thing for the Aivilik to believe in is death, an
episode on the road of the immortal life of man. One mustn't
make too much of life on earth. An Aivilik death chant begins,
"Say, tell me now, was life so nice on earth?" the answer being,
of course, that it wasn't, and that life hereafter is likely to be
much nicer. There is hope this side of despair.

It is often erroneously assumed that because nonliterates respond to the presence of death with great stoicism they are
wanting in feeling. This is far from being the case. In the first
place, the expression of emotion in the presence of death takes
different forms in different cultures. In some, as among the
Australian aborigines, it is permitted in the form of wailing and
mourning ceremonies, but in others, as among many American
Indian peoples, the expression of emotion in such ways is
strongly discouraged.

Deep emotion is no less felt by American Indians than it is by

Australian aborigines; it is its expression which is simply differently controlled. At one funeral ceremony, a Winnebago Indian rose and addressed the visitors as follows:

It is said that we should not weep aloud and you will, therefore, not hear any of us making any utterings of sorrow. And even although we weep silently we shall smile upon all those who look at us. We beg of you all, consequently, that should you find us happy in mood, not to think any the worse of it.

Elsdon Best relates that when an old Maori chief was suddenly stricken ill and his aged wife began to lament for him, he calmed her with the words, "Do not lament. It is well. We have trodden the path of life together in fair weather and beneath clouded skies. There is no cause for grief. I do but go forward to explore the path." Such thoughts are enshrined in innumerable beautiful chants and poems composed by every people in response to deeply felt emotion. The Maori of New Zealand cherish the thought just quoted in the following poem:

> The tide of life glides swiftly past
> And mingles all in one great eddying foam.
> O heaven now sleeping! Rouse thee, rise to power;
> And thou, O earth, awake, exert thy might for me
> And open wide the door to my last home,
> Where calm and quiet rest awaits me in the sky.

Hand in hand with the largely objective view of death that nonliterates take is the objectivity of their view of life—the tragic view of life, the sense that life is "not all beer and skittles," that tragedy is a part of the experience of human life, and that man must reconcile himself to the inevitable. Different nonliterate peoples have adjusted themselves to this fact in varying ways, some attributing the tragedy in life to man's own transgressions, while others attribute it to the very nature of things. The adjustment is very generally a realistic and sensible one. Some individuals take the tragic element in life a little harder than others, while some settle for regarding life as tragic in part, but also in part comic—a tragicomedy. Wisdom, then, lies in coming to terms with the facts of life. Since it is inevitable that human beings will suffer losses, deprivation, and injustice, that they will come into conflict with some, at least, of their fellows, and even loved ones, and with society, and that much suffering will ensue, it is best to be prepared. Though suffering is inevitable, it cannot usually be foreseen; one cannot control or alter it. It is from this feeling of helplessness that the tragic sense of life arises. But clouds have silver linings, as is so well brought out in the following Papago Indian "Song to Pull Down the Clouds":

> At the edge of the world
> It is growing light,
> Up rears the light.
> Just yonder the day dawns,
> Spreading over the night.

Speculation for its own sake occurs among all nonliterate peoples, though it is not to be thought that this is a habitual practice with everyone in the group, any more than it is with all the members of civilized populations. Radin gives this remarkable example, which must do service here for many others, of a Plains Oglala Indian's attempt to explain the significance of the circle for his people:

The Oglala believe the circle to be sacred because the great spirit caused everything in nature to be round except stone. Stone is the implement of destruction. The sun and the sky, the earth and the moon, are round like a shield, though the sky is deep like a bowl. Everything that breathes is round like the body of a man. Everything that grows from the ground is round like the stem of a plant. Since the great spirit has caused everything to be round mankind should look upon the circle as sacred for it is the symbol of all things in nature except stone. It is also the symbol of the circle that marks the edge of the world and therefore of the four winds that travel there. Consequently it is also the symbol of the year. The day, the night, and the moon go in a circle above the sky. Therefore the circle is a symbol of these divisions of time and hence the symbol of all time.

For these reasons the Oglala make their *tipis* circular, their camp-circle circular, and sit in a circle in all ceremonies. The circle is also the symbol of the *tipi* and of shelter. If one makes a circle for an ornament and it is not divided in any way, it should be understood as the symbol of the world and of time.

In this remarkable passage we perceive the functioning of the human mind at very nearly its highest level: the imaginative systematization of ideas, based upon what to others would be, for the most part, a series of bewildering experiences.

With respect to the systematization of ideas, nonliterate peoples have distinguished achievements to their credit. The Australian aborigines' creation beliefs constitute a complete evolutionary theory, foreshadowing the evolutionary theories which were later worked out by scientists. As we have already learned, the aborigines postulated a time when there were no living things. These were then created by two formless spirits who later disappeared. There were at first no plants or animals, but only the precursors of these; then gradually these half-plants and half-animals, as they call them, were converted into real plants and animals, and from the latter men were eventually evolved. It is for these reasons that the aboriginal considers himself related to all natural things.

There are many similar stories of creation in different cultures. All of them represent man's attempt to explain the world to himself.

❧ 14 ❧

SCIENCE

Wʜᴀᴛ ɪs sᴄɪᴇɴᴄᴇ? The Nobel Prize-winning physicist Percy W. Bridgman has described it as intelligence in action with no holds barred. It is doing one's utmost with one's mind. And that, of course, is what most scientists try to do. If that is what makes one a scientist, then many scientists have appeared among nonliterate peoples. But, of course, such a description of science is a bit too general, for doing one's utmost writing a story or cutting a better figure at ice-skating doesn't necessarily make those activities scientific. It is therefore necessary to arrive at a more accurate description of science. Science is (1) a form of behavior characterized by hard thinking and testing, designed to discover what *is,* and (2) the systematized knowledge or system of facts and principles concerning any subject, arrived at in a verifiable manner. Much more than the facts, science is a system of concepts, the function of the facts being to test and correct the concepts. Science must be distinguished from the mere application of rules and methods, which is *technics.* Scientific procedures have usually been followed in arriving at those rules and methods, but it is the intellectual procedures that have been used and not the rules and methods themselves that constitute science. Science, in short, is the effort of the human mind to enlighten itself by evidence.

A laboratory technician is not a scientist. He is a wielder of methods. The person who has devised the ideas which are implemented by the methods the technician is using is the scientist. In this sense there are scientists as well as technicians in nonliterate cultures. In fact, in most nonliterate cultures everyone is a technician in one or more materials, but scientists are no more frequent than they are in any peasant society of the Western world.

Consider the following: Suppose people were born into a culture where no one knew how to make fire, where people had seen forest fires but were scared to death of them. How long do you think it would take to discover how to make fire and put it to use? The fact is that it might take a very long time, hundreds of thousands of years perhaps. The extinct Tasmanians are said not to have known how to make fire, but to have obtained it from fires which they constantly kept going. When Professor A. R. Radcliffe-Brown studied the Andaman Islanders during the years 1906 through 1908, he discovered that they knew of no way to make fire other than by kindling it from another ignited piece of wood. Fires were carefully kept alive in the village and as carefully carried when traveling. The Andamanese showed great skill in selecting wood that would smolder for a long time without going out and without breaking into a flame.

The Pygmies of the Epilu River of Central Africa have not learned, or they refuse, to make fire. They purchase it on the barter system from their Congo Negro neighbors.

In order to put fire to use, one must first *predict* a use for it. One must see ahead clearly what it will be able to do; in other words, a theory must be developed to the effect that if one could capture and control fire one might be able to make it do one's bidding. Thinking of this kind is the essence of science. For example, as far back as the fourth century B.C. Democritus (460?–370? B.C.) suggested that matter was made up of tiny particles, imperceptible to the senses, which he called atoms. This was a theory. It took over two thousand years to prove that theory correct by the demonstration of the actual existence of the atom. The man who first thought of capturing fire for human use was every bit as great a genius as was Democritus. He is one among the thousands of unsung benefactors of the human race. Prometheus is the name the ancient Greeks gave to the Titan who first taught man the use of fire. The word "Prometheus" means "forethought," and forethought, the last of the gifts granted by the gods to man, is the father of invention. The tale is that he made man of clay, and, in order to endow the clay with life, stole fire from heaven and brought it to earth in a hollow tube. Similar tales come from many cultures other than the Greek.

Having, then, with forethought foreseen the value of controlled fire, the next step was to capture it. How this was achieved we do not know. There are many theories but none can be proved. One theory suggests itself with great force, and this is based on the coincident appearance of flint-chipping and fire. The suggestion is that in chipping his flints, prehistoric man could not have failed to observe the similarity of the sparks to the fire induced from natural causes—spontaneous combustion, lightning-struck trees, prairie and forest fires, perhaps the rubbing of dry twigs against each other in a high wind, or maybe fires induced by volcanic action. Furthermore, such sparks may have ignited dry grass, and in this manner led to the accidental discovery of fire-making. It may be that fire was independently discovered by different groups of men in different ways. We do not know. But what we do know is that someone did test the theory that if one imitated the conditions under which fire was naturally produced, the probability was that one could produce it artificially. This was a piece of scientific reasoning and experiment, and it succeeded. Methods of making fire vary in different nonliterate groups. The earliest known use of fire by man is associated with Peking man (*Homo erectus pekinensis*), but how he made his fire, or whether he simply used it as the Andamanese do, we do not know.

METHODS OF MAKING FIRE

Since the variety of methods of making fire illustrate the scientific uses of the imagination, they may briefly be described here. Small nodules of iron sulfide (iron pyrites) are of common occurrence. When two pieces are accidentally knocked together, they give off sparks. The forcible striking of two sulfides, or one sulfide and a piece of flint, chert, quartz, chalcedony, or other siliceous stone, is the method of *percussion*. The evidence points to the fact that in the Old Stone Age, fire was made by the use of flint and pyrites—a method still in use among the Eskimos, the Fuegians, and some American Indian tribes. Tinder in the form of dried moss or fungus, the down of floss or seeds, and dry and rotten wood seems to have been discovered early and used widely. Astonishing as it may seem to those of us who are unacquainted with the fact, sparks can be struck from the most unpromising materials. For example, in southeastern Asia—the Malay Peninsula and the Philippines—a rough-surfaced bamboo is struck with a piece of porcelain to produce sparks.

FIRE BY WOOD FRICTION

By the friction of one piece of wood upon another, wood dust is produced. During the friction so much heat is generated that the little heap of wood dust begins to smolder, and in a few seconds can be blown into a glowing flame. Fire may thus be produced in a few minutes or less. There are three principal primitive tools used for making fire by wood friction: (1) the saw, (2) the plow, and (3) the drill. The wood that is held at rest on the ground is called the *hearth*. One either saws across the grain of the hearth, plows along it, or drills into it at right angles. The last method is the most widely distributed.

THE FIRE-PISTON

The fire-piston is a very simple instrument which must have taken, however, the acutest observation and intelligence to devise. Its fire-making depends upon the fact that if air in a confined space is abruptly compressed, heat is developed, which may then set fire to tinder. It is widely distributed through southeastern Asia. It consists of a narrow cylinder four to six inches long of bamboo, wood, or horn, and sometimes metal, closed at one end. Fitted tightly into the cylinder is a piston of similar material. At the lower end of the piston is a small depression in which the tinder is lodged. The piston, drawn to the top of the cylinder, is given a smart push to the bottom of the chamber, and on withdrawal the tinder is found to be alight. The instrument seems to have been first invented in Asia, and was not reinvented in Europe until the beginning of the nineteenth century, where it went out of use in 1827 with the introduction of the friction match.

The discovery of the fire-piston was almost certainly due to an accident of favorable observation by an acute intelligence. It probably did not involve the working out of the principle first and its application afterward. Nevertheless, the original discoverer was making quite as significant a scientific finding as

Fire-piston from southeast Asia

was Sir Alexander Fleming when he observed in 1928 the effects of the penicillin mold upon his plate of bacteria.

One must make a sharp distinction between discovery by *finding*—a uranium deposit, a new star, or an Indian burial ground—and *discovery* by finding out. The latter involves the uncovering of the ways in which things work, their properties, and the conditions which relate events to each other, and finally it involves applying this knowledge to the accomplishment of a purpose. Observation, discovery, and application are the three processes involved. There can be little doubt that early man and nonliterate man of later times, to the extent that they indulged in scientific activities, did so in an intensely practical way and for exclusively practical purposes.

Agriculture was a scientific discovery of great practical value arising out of practical needs. The invention of many of the implements devised to assist the agricultural process, and many of the techniques used in planting, involved scientific thought. Practical or applied science is an accomplishment of every non-literate people. It differs from that of more sophisticated societies in that it is generally applied on a smaller scale and to the solution of a comparatively limited number of problems.

TIME-RECKONING

Of practical importance to some nonliterate peoples is a method of reckoning time in order that the coming and the going of the seasons may be predicted, the waxing and waning of the moon, the migration of animals, the rise and fall of rivers, and so on. Every nonliterate people has some method of reckoning time, though among some the method is much more crude than among others. The simplest way to reckon time is to note the relation between the appearance of one event and another which follows it more or less regularly. When certain plants blossom, for example, certain kinds of animals also begin to make their appearance and others will follow. The seasons are everywhere recognized, made use of, and subdivided.

The Greenland Eskimos divide their seasons in correspondence with the wanderings of animals, the position of the sun, moon, and stars, and other observations based on the regularities of their environment. Most nonliterate peoples have a calendar, however crude in some cases, which is based on astronomical observations. It should be remembered that most nonliterates sleep out in the open, and during the night are able to become more thoroughly acquainted with the stars in the heavens than most civilized men ever do. There are few, if any, nonliterate peoples who do not have names for the larger stars and star clusters. The temporary appearance and disappearance of the stars is not a matter of random or occasional observation with nonliterate peoples. They are therefore easily able to associate other changing natural events, such as the changing seasons, with the stellar changes, and thus use the stars, the sun, and the moon to predict coming events. Along with predicting events, they are able to count the time between the intervals.

The position of the sun, and the type of shadow it casts upon any object at particular times, in association with certain natural events, drew the attention of many different nonliterate peoples. They observed that the sun is in the north during one season and in the south during another (the two solstices), and that the shadows cast upon the same object in the same place vary not only with the hour but during different days of different

seasons. The Australian aborigines have made excellent use of
this observation.

The Australian aboriginal has a good idea of time. In referring to a past or future event, he will describe it by pointing to
the assumed altitude of the sun. To fix a time for a proposed
action, a stone is placed upon a cliff or in the fork of a tree, a
day or two in advance, so that the sun strikes it exactly at the
hour decided upon. When this occurs on the day of the proposed action, those involved know that it is the time agreed
upon. Days are reckoned by the number of sleeps, and larger
stretches of time are counted by the number of moons.

The Zuñi Indians of the American Southwest determine the
beginning of their new year, and the time for planting and
harvesting, by using a large, upright block of sandstone as a
datum point for taking observations of the sun.

Volumes could be filled on the astronomical observations and
the measurement of time made by nonliterate peoples. The examples cited here serve to illustrate the fact that nonliterate
peoples do, scientifically, attempt to measure the passage of
time in order to bring the world under better control. We cannot, however, complete this section without some reference to
those remarkable Vikings of the Sunrise, the Polynesians, who
read the heavens and the tides so well that they were able to
draw up the most elaborate and detailed guides to navigation.
These navigation charts were remarkably accurate and helpful,
and made the Polynesians the greatest navigators of their time
—up to and even after Columbus.

NUMBERS AND MEASUREMENT

The idea of number probably occurred very early in the history
of man. The food-gathering peoples seem to have the least need
for numbers in their daily lives, and therefore conditions prevailing among contemporary nonliterate food-gathering peoples
probably closely resemble the state of mathematical knowledge
of early man.

The Australian aborigines vary in their methods of numeration. In general they have no running series of numerals. The
Arunta of Central Australia occasionally counted up to five,

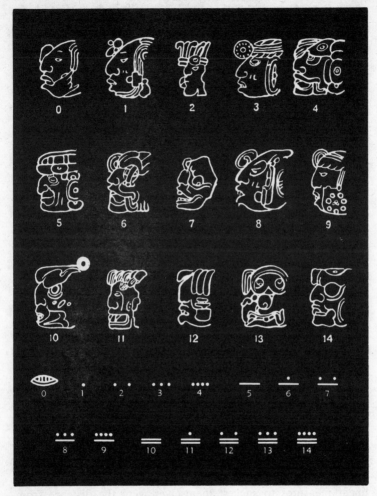

Mayan. Two types of numerical notation

sometimes using their fingers in doing so, but frequently anything beyond four was indicated by the word *oknirra,* meaning
much or great. One is *ninta,* two *terama* or *tera,* three *tera-ma-
ninta,* four *tera-ma-tera,* five *tera-ma-tera-ninta.* There are words
for "small number," "larger number," and "very large number."
Slightly, but only slightly, more elaborate systems of counting
than this are found among some tribes, such as the Dieri tribe

near Lake Eyre in south-central Australia, who count in the same way as the Arunta, but up to eleven, bringing in the feet and toenails as well as the hands. The fact is that food-gathering peoples seem to have very little use for complex systems of numeration and therefore haven't developed any. It is when populations become large and the culture more complex that the necessary mathematics is developed to meet new needs.

The Maya of Yucatan not only invented methods of addition and subtraction, but they also invented a symbol for zero and gave a value to their numbers according to zero's position. The Aztecs indicated 1 by a finger, 20 by a flag, 400 by hair, 8000 by a pouch.

There is evidence that in addition to their other remarkable architectural achievements the Maya also independently invented the arch, which is found nowhere else in the Americas except at Chichen Itza.

Beam scales were invented by the Peruvians together with an original method of determining equilibrium.

In nonliterate societies there is little need for precise forms of linear measurement. Where measurement, as among the Kwakiutl of Vancouver Island, becomes essential in laying out the lines for a square house, considerable precision has been attained in the application of geometric principles. This is seen also in their manner of making boxes.

Such a thing as the objective standardization of measurements is rarely found among nonliterates, though a quasi-objectivity in linear measurements is reached in the widespread use that has been made of the proportions of man's own limbs. The length of the forearm from elbow point to tip of middle finger is a good primary unit for laying off lengths. Obvious linear measures to suggest themselves were the distance from the outstretched thumb tip to the tip of the outstretched little finger, roughly equal to half the forearm length, and the length from outstretched forefinger tip to little finger tip (the little span), equal to one third of the forearm. Others were the foot length, palm length, hand width (a horse still stands so many "hands" in height), and finger width (still a measure in modern medicine), and then, of course, there was the height of a man. The forearm length became the cubit, and four cubits made a fathom.

Notched sticks for making more or less exact measurements

are in quite wide use among nonliterates, and cloth and other materials are often measured out by their use. Such sticks were also used in building boats by the Penobscot Indians of Maine.

For the measurement of volume, all sorts of familiar containers are used, from shells to baskets and canoes. Let it be recalled that the oilmen still measure the yield of their wells and the content of their reservoirs by the barrel, a reminder of the fact that oil, by sheer accident, happened to be transported first in barrels.

The techniques of smelting, extracting, and alloying ores involved many a scientific process, but these techniques were not really developed until about 3500 B.C. in the Near East. Copper, gold, and meteoric iron are surface ores found in relatively pure form, so that they require no smelting to make them usable. The Neolithic Badarians of Egypt were able to cold-hammer such ores into tools and ornaments, and several American Indian tribes that lived where surface outcroppings of copper were available did likewise. The Eskimos use bits of meteoric iron to make tools. Using these metals, the Aztecs and the Incas excelled in fine work.

In 1964, in the region of Diyarbakir, near the Tigris headwaters in southeastern Turkey, cold-hammered copper pins were discovered at a site believed to be some 9000 years old, which would make them the earliest copper tools known.

Ironworking is widespread, as is the smelting of iron ore, among African peoples, who may well have discovered the processes involved in forging, extracting, smelting, and manufacturing implements of all sorts. It is, however, believed by some authorities that the ironworking of the Africans, like that of the Europeans and Asiatics, diffused to them from the Near East. The oldest African datable iron tools are from Meroë in Nubia (700 B.C.).

INDEPENDENT INVENTION

Under similar challenging conditions the human mind everywhere tends to respond in similar ways. One would therefore expect that similar inventions have been independently made many times by peoples widely separated from one another, but not altogether surprisingly it is very difficult to produce more than a few instances of probable independent invention. These are the blowgun shooting arrows tipped with poison in the Malay Peninsula and the Amazon Valley, the symbol for zero in the case of the Maya and the Asiatic Indians, the suspension-cantilever bridge of New Guinea and the Western world, infibulation in Peru and in Africa, and the bull-roarer, a flat piece of wood which makes a buzzing sound when spun around at the end of a cord, which occurs in Australia, Melanesia, Africa, and South America.

The difficulty of proving independent invention does not mean that it has seldom occurred, as some extreme *diffusionists* have held, but simply that though it must have often occurred it is difficult to prove. *Diffusion* is the process by which cultural traits *have been* transmitted from one group to another—to be distinguished from *acculturation,* the process during which cultural traits *are actively* being transmitted from one people to another.

Enough has been said to indicate that nonliterate peoples are capable, under the pressing conditions of necessity, of doing their utmost with their minds in solving practical problems in a scientific manner. Most of the scientific processes involved are of a practical nature, and there is little time or inclination for science for science's sake. The latter activity does not appear until the development of highly sophisticated societies like Hellenic Greece. This is but yesterday in the history of human time. Because the Greeks were an aristocratic society based on slavery, they, who had so much of the requisite theoretical knowledge at their disposal, largely failed to apply it. Machines were unnecessary since slaves could do all the work. The Greeks were interested in ideas—in brains—not in drains. The refuse of civilization could be disposed of by slaves, but only those with the necessary leisure could create and maintain that civilization. The Greeks developed the greatest ideas in the humanities and the sciences that the world has ever known, and probably the fewest inventions of a mechanical kind. Not that the Greeks were uninterested in the practical application of some of their ideas; it is simply that they were not enamored of the possibility of creating machines that think and human beings who don't. With that, they clasp hands with the long millennia of nonliterate men.

➷ 15 ➹

THE ARTS

THOSE PURSUITS in which the imagination is chiefly engaged, the giving of form and meaning to the materials with which one works, are known as arts. In this sense the arts are probably very old. There is no known people that is without them—drawing, painting, carving, sculpture, music, poetry, storytelling, and the dance.

To draw seems to be a natural propensity which children seek to express at the earliest possible opportunity. Without any instruction whatever, small children will spontaneously make drawings with the materials placed in their hands. Apes will, too. Chimpanzees provided with brush and paint delight in producing "nonobjective art." Dr. Desmond Morris of the London Zoological Society has written a book describing the "art" of a particularly gifted chimpanzee. At least one young gorilla has actually been observed tracing his own shadow with his forefinger. This is not unrelated to the way children draw. One little girl when asked what she did when she drew replied, "First I think, and then I draw a line around my think."

Drawing in all its forms constitutes one of the languages of the imagination. Since it is the oldest form of art of which we have any record, perhaps we had better begin with it.

DRAWING AND PAINTING

Ninety years ago if anyone had suggested that prehistoric men of the Old Stone Age were at least as intelligent as their modern descendants, he would have been greeted with derision. If he had gone on to suggest that there have been artists who lived some fifteen thousand years ago whose artistic accomplishments could scarcely be bettered by any artist who has lived since, his sanity might well have been questioned. And yet these statements are perfectly true, for the oldest known paintings by man go back well over ten thousand years, and they display a really extraordinary skill and beauty. As Sir Herbert Read, the distinguished English art critic, wrote:

The mystery—for it is still a mystery—is why, at such a stage of human development, the art of representation could reach a degree of refinement that has never subsequently been excelled. Such a statement may be challenged, but allowing for the coarseness of the surfaces on which the prehistoric artists painted, and the primitiveness of their tools and colours, the degree of skill exhibited in the best drawings in the Altamira, Niaux, and Lascaux caves is not less than that, say, of Pisanello or Picasso.

The first discovery of prehistoric painting was made in 1879 near the village of Santillana del Mar about twenty miles west of Santander, Spain. Here, on the ceiling of a large cave that was named Altamira, paintings and drawings were discovered of such astonishing competence that for almost thirty years the experts refused to accept them as the work of prehistoric man. It was not until many more caves had been discovered showing the same kind of drawings and polychromes that the experts were at last forced to yield to the facts.

Often these drawings and paintings are found in the innermost recesses and darkest and most uncomfortable corners of the cave, and often, too, they show one drawing or painting superimposed upon another. The animals shown are mainly of extinct species in the southwestern European region—bison, wild boar, wild horse, mammoth, reindeer, ibex, antelope, woolly rhinoceros, cave lion, moose, musk ox, and several others. From the fact that these animals are depicted and that their fossil remains occur in the caves, associated with the hearths of the human inhabitants, it has been calculated, by radiocarbon dating, that the cave paintings made at Lascaux in central France (discovered in 1940) were executed about fifteen thousand years ago. This happens to be almost exactly the estimate that archeologists had arrived at by other means.

It is interesting to follow the distribution of this cave art from the Dordogne of central France southwest to Spain, for it sug-

Example of Upper
Paleolithic art

gests a group of populations who were in touch with one another, and who clearly exhibit a number of cultural traits in
common, of which their art was but one.

Why did these people of Aurignacian culture choose to do
their painting in uninhabitable caves, and in the darkest and
most inaccessible recesses of these? The most likely answer is
that their paintings were made for magical purposes and not in
order to decorate the caves. The animals shown on the roofs
and walls of these caves are often represented as pierced by
spears and arrows. One creates as naturalistic a model as one
can of the animal one hopes to kill, and then kills it in effigy; as

Presumed sketch sheet (lower) and final cave painting (upper)

one does to the drawing of the animal—accompanied by the proper ritual—so one will do in fact to the real animal. One has but to wish in the ritually acceptable manner and one will succeed. Hunting scenes abound in these cave drawings and paintings, and there can be very little doubt that cave art, at any rate, was devoted principally to the practical purpose of securing success in the hunt. This does not mean that the artist did not obtain some esthetic pleasure from his achievement, but it does mean that love of beauty was not his principal motivation.

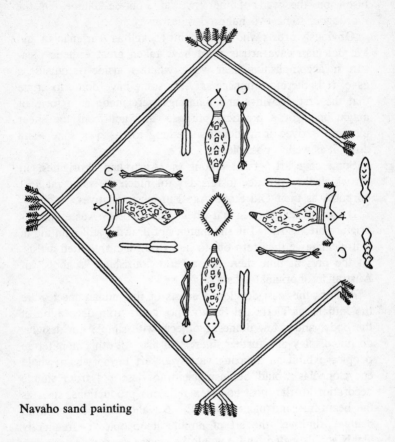

Navaho sand painting

That the artist took pride in his accomplishment and was encouraged to do so is indicated by several facts. In the first place, the skill exhibited by these artists takes a certain amount of training. That such training was available from different centers is testified to by the fact that preliminary sketches on stone of certain animals have been found several hundred miles away from the caves in which the finished polychromes occur. In 1926 an engraving on limestone was found in the half-cave at Genière in France. This was immediately recognized to be the sketch for the very individually painted polychrome of an old bison on the wall of the cave at Font-de-Gaume, in the Dordogne, some two hundred miles away.

Obviously artists who could paint as well as Aurignacian and Magdalenian cave artists must have taken great esthetic pleasure in accomplishing their work, whether inside or outside a cave. It is therefore unnecessary, as some have done, to argue that the first drawings and paintings were made as a form of magic only for a practical purpose, and not from the sheer pleasure derived from doing something for its own sake—and doing it as well as possible.

Some cave art persists among the Australian aborigines, an art which they practice to this day for much the same magical purposes as their Old Stone Age European relatives.

It is of interest to note that when contemporary Australian aborigines are instructed in the manner of drawing and painting in European style they turn out to be extraordinarily gifted artists. At the present time there are several flourishing "schools" of Australian aboriginal artists.

Perhaps the least developed artists of the human species are the Indians of Tierra del Fuego who, apart from decoration of the body, and a few domestic objects with simple dot designs, do not display any further interest in art. All other nonliterate peoples exhibit an amazing variety of art forms which whole encyclopedias would be insufficient to describe, from simple decoration to the most complex religious productions, such as the beautiful sand paintings of the Navaho Indians. Far from being "primitive" the art of nonliterate peoples is frequently highly sophisticated, and it would be nearer the truth to say that in comparison the art of literate peoples was, until very recently, "primitive" in the sense of being more representational and very little evolved. It is to an appreciable extent due to the discovery of "primitive art" in the twentieth century that literate art has at last developed away from the naturalism which has for so long characterized it.

Whether it be practical, magical, religious, or decorative, the function which art serves in nonliterate cultures, if it can be said to serve a single function, is to relate the individual to the group. Whatever is represented is caculated to interest and bring the group closer together in some way.

LITERATURE

The unwritten literature, poetry and prose, of nonliterate peoples is often very highly developed. This literature is known to us only in translation, but those who are familiar with the native rhythms, if not with the languages, tell us that they are quite as beautiful as any to be found in the literatures of the Western world.

The food-gathering peoples have a literature which matches that of any more advanced people. Every tribe seems to have its poets and storytellers and their gifts are much appreciated in nonliterate societies. In reading the recorded literature of nonliterates, it should always be remembered that it is in translation and that its native character has probably been profoundly changed by the Europeanization of the rhythms and even of the meanings. The excellence of the original can therefore rarely be fully appreciated. The best work of its kind which preserves the spirit of the original as nearly as possible is Frank Hamilton Cushing's *Zuñi Folk Tales* (1901), and Verrier Elwin has done the same for the native peoples of Chhattisgarh, in the Central Provinces of India, in his *Folk-Songs of Chhattisgarh* (1946).

The theme of literature everywhere is life and death and all that happens in between. Every literature represents an expression of the culture. Among nonliterates both poetry and prose are closely associated with song, and most of the poems and even stories are either sung or accompanied by song. Rhythm in poetry and prose, as well as repetition, which is a form of rhythm, is markedly emphasized, and it is no accident that poetry, prosody, song, instrumental music, and the dance are all closely associated. The poetry and prose of the Old and New Testaments retain the devices of rhythm and repetition to a marked degree, traits which are characteristic of Near Eastern literature as a whole, and which are undoubtedly among the most ancient of all literary traits. Sometimes the repetition may be extended to great length, and frequently, as in the following quatrain, used with great and ingenious skill. The poem is a Nootka Indian one and is entitled "Plaint Against the Fog":

Don't you ever,
You up in the sky,
Don't you ever get tired
Of having the clouds between you and us?

As among most nonliterates, the Copper Eskimos of Victoria
Island, just south of the North Magnetic Pole, highly value those
who are able to create new songs as well as dances, and almost
every one of them possesses this talent. Here is a Dance Song:

I am quite unable
To capture seals as they do, I am quite unable.
Animals with blubber since I do not know how to capture,
To capture seals as they do, I am quite unable.
I am quite unable
To shoot as they do, I am quite unable.
I am quite unable
A fine kayak such as they have I am quite unable to obtain.
Animals that have fawns since I cannot obtain them,
A fine kayak such as they have I am unable to obtain.
I am quite unable
To capture fish as they do, I am quite unable.
Small fish since I cannot capture them,
To capture fish as they do, I am quite unable.
I am quite unable
To dance as they do, I am quite unable.
Dance songs since I do not know them at all,
To dance as they do, I am quite unable.
I am quite unable to be swift-footed as they are,
I am quite unable . . .

Among the Copper Eskimos the dance song takes the place
of the daily newspaper, for in the great Copper House where the
Eskimos daily assemble, every important incident is recorded in
dance song. Song and dance, poem and story serve not only to
assist in binding the group together, in keeping it informed and
alive by serving as the vehicles for traditional lore, for myth,
ballad, and magic formula, but also serve as a means for reduc-
ing personal and interpersonal tensions. For these purposes
many peoples provide opportunities for dramatic experiences,
ceremonies, and plays, in which every member may play a part.
That such experiences can have a beneficial effect was discovered by nonliterate peoples long before the development of
modern psychology.

MUSIC

All peoples sing, but many nonliterate peoples do not possess any musical instruments except those used for expressing rhythm. When this is true, tone is provided by the human voice. Sometimes the range of melodies is limited to a few notes, and sometimes it extends over more than an octave.

Vocal music probably developed before instrumental music, and in all nonliterate societies the characteristics of *direction* and *interval* are found. Direction, which may be either tumbling or horizontal, derives from two distinct psychological roots of singing, the *tumbling* type representing the emotional expression of tension commencing with a high note, and after bounding upward then descending, and the *horizontal* type proceeding as a calm recitation, sometimes of only two pitches sung in alternation. Both types of song may be customary in a single group and may even form part of the same song.

The melodic steps of the voice, *intervals,* occur in great variety. Melodies may be limited to one or two notes and tones, like the first babble songs of children, or consist of seconds or thirds or fourths, and so on. For example, while we divide the octave into twelve equal intervals, the Siamese divide theirs into seven, the Malaysians generally into five equidistant steps, and American Indians employ both thirds and fourths in the intervallic structure of their melodies. The intervals employed often differ markedly from the tempered scale to which we are accustomed, and sound strange to our ears. But when we come to appreciate such music we find that it has a dignity, sincerity, and beauty all of its own.

The comparative study of the music of different peoples is an important branch of anthropology, and is known as *ethnomusicology*.

The Australian aborigines sometimes use a tubular conch shell as a trumpet, but merely to enlarge the sound of the human voice. Reeds are also commonly used, and sticks, both to produce rhythm rather than tonality. The Bushmen and Hottentots use the musical bow and a reed pipe, and they sometimes fashion crude flutes. The elder men teach the young boys

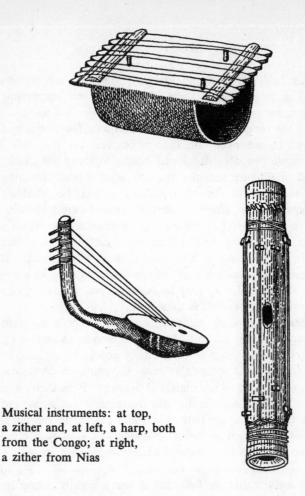

Musical instruments: at top,
a zither and, at left, a harp, both
from the Congo; at right,
a zither from Nias

to play these instruments, and girls look with much favor upon
the boys who excel. Drums and all sorts of other percussion
instruments are widely used among nonliterate peoples. Gongs
and the xylophone are used for melodic purposes, the former in
southern Asia, the latter in Africa. Notched bones or sticks
pulled over hollow vessels yield a rasping sound, and such
devices have been used by many peoples. There are all sorts of
original instruments, never used by us, which were invented by
nonliterates. One is made of a block of wood which has been

divided by deep curved notches into a number of sections. By striking the sections, or rubbing over them, different tones may be produced. This instrument comes from New Ireland in Melanesia. The African *zanza* consists of iron strips of different lengths stretched over a resonating box provided with a bridge. The strips are pulled at their ends like strings. String instruments seem to have been quite unknown in the Americas before the advent of the white man. In Africa they are widespread, the harp and the lyre being among the most frequent in the African world. The banjo is an African instrument.

THE DANCE

The dance is almost certainly as ancient as song. Singing appears to be spontaneously accompanied by rhythmical movements of the body. Many of the cave drawings and paintings show unmistakable representations of various dances—women dancing around a man, a medicine man dancing, and the group as a whole dancing. Then there are various small groups dancing together. All these varieties of the dance occur among nonliterate peoples today, and they occur also among the more advanced societies. What is rare among nonliterate societies are joint dances where single couples stay together throughout the dance as among ourselves. Nonliterates dance all together; we tend to dance in pairs, with each couple dancing alone, not with the other couples. Furthermore, in the dancing of nonliterate peoples continuous body contact between the sexes is extremely rare.

As is true in almost everything else, the dance among nonliterate peoples is a group phenomenon, in which everyone actually or vicariously indulges. Occasionally a particular dance may be the property of a particular person. In New Ireland anyone who becomes enamored of a dance may, if the owner is willing, purchase it. In Africa even a particular step may be the property of some organization.

Dances may be indulged in for the purposes of amusement, or for ceremonial, religious, or esthetic reasons. Whatever the motive, every culture has its own dances, and they are often of the greatest beauty, and even the least beautiful is likely to be interesting. The universal need for expression in rhythmical bodily movements is differently satisfied in different cultures, according to the stylized ways customary in each. But everywhere in nonliterate cultures the function of the dance is the same: to give expression to a need and to bind the dancer closer to his group. There is dancing for the sheer joy of it also, but even the form of this kind of dancing is largely determined by the group.

TRANSFERENCE OR COMBINATION

A common cultural process, affecting other activities as well as the arts, which is observable in all cultures, is *transference* or *combination*. Various components are taken from different things and transferred to another thing, where they are combined in a novel way. For example, a design may be taken from a basket and transferred to a pot, which is itself a movable container originating in a cooking hole—a hole as it were together with its earthen walls lifted out of the ground. A plow is a combination of digging implement and a source of energy hitched to it, either a man or a domestic animal. Elements of a myth may be borrowed from another people and incorporated on one's own myths. Dances, like the Sun Dance of the Plains Indians, may be created from elements taken from those of surrounding tribes.

❧ 16 ❧

THE ANTHROPOLOGIST
IN THE FIELD

THE OBSERVATIONS DESCRIBED in the preceding pages are largely those made by anthropologists working in the field. These observations represent, for the most part, a minute fraction of the labors of many investigators who have been at work during this century. One of the first lessons to be learned from reading a short book such as this, which attempts to deal with so large a subject, is that, far from coming to the end of a book, one comes, in reality, only to the beginning of many others. The bibliography, which follows this chapter, lists a number of books, briefly annotated, which is designed to introduce the reader to the larger literature relating to the various aspects of anthropology.

In the first chapter of this book, the reader was briefly introduced to anthropology. In this final chapter, I should like to bring the reader up to date on what anthropologists are actually doing at the present time.

Nonliterate cultures are disappearing or changing at so rapid a rate that there is not a minute to be lost in studying them. Cultures that have remained untouched by civilization still exist, but as time goes on they will become fewer and fewer. Anthropologists have long realized this and have done all they could to have such cultures studied. Usually the anthropologist will have studied one or more nonliterate cultures in the field, and in training his students will require of them that they apply the theories and the techniques they have learned to the study and monographing of at least one nonliterate culture. The importance of such work cannot be overestimated. In the first place, insofar as the anthropologist himself is concerned, fieldwork is a necessary part of his training. Just as the internship is considered a necessary part of the training of a physician, so today internship in the field is considered an indispensable part of the training of every anthropologist. The anthropologist who has not had the advantage of such field experience may achieve high competence as a theoretician in the particular aspects of anthropology that he has made his own and may as a practical anthropologist be indistinguishable from the best, but he will usually be lacking in a certain "feel" for, and a certain insight into, the nature and meaning of the cultures that have been studied by other anthropologists. There are, of course, exceptions to every rule, but there can be no question of the value of having undergone the experience of living with a people whose way of life differs so substantially from that in which any anthropologist is likely to have been socialized. Learning to see another people's way of life, its culture, objectively is in itself an invaluable experience. This is not only what the anthropologist tries to do himself, but also what he tries to teach other people to do.

In studying the manner in which the other half of the world conducts its affairs, we not only learn how remarkably various man's adaptations and adjustments to his environment can be, but we also learn much about the malleability of human nature. Not only that, we learn a great deal about human nature itself. Indeed, it is anthropology, and not psychology, which has been able to show how unsound is the old saying, "You can't change human nature." On the contrary, anthropology has illuminatingly and convincingly shown how changeable human nature is. The usual concomitant of the view that one cannot change human nature is that if one is to embark upon any form of social change without creating havoc, it must be done very gradually and over a long period of time. Recent studies—such as those of Margaret Mead in the Admiralty Islands, where the Manus have made the passage from the Stone Age into the Jet Air Age within a few years, with anything but catastrophic effects—indicate that the rate of social change can proceed at quite an accelerated pace under reasonably favorable conditions.

In the United States one often hears it said that in connection with the changing status of the Negro it is necessary to "go slow." This is often a rationalization for not going at all, for it may well be asked, How slow? Were there the will the long overdue changes could be made with great speed, but the divisive racism, which in reality represents the repudiation of humanity, stands in the way. The President and the Congress have explicitly guaranteed the equality declared by the Constitution as the right of every citizen, and this has been repeatedly affirmed by the highest courts. Nevertheless, and in the face of protests by vast numbers of Negro citizens and others across the length and breadth of the land impressing the urgency of their need upon the nation, at the present only token equality has been achieved in the United States. Racism which began as a canard, may yet end in catastrophe. The pattern of conformity is closely related to the pattern of exclusiveness. Racism is irrational, but that does not mean that it cannot be influenced by rational means. The hope is that reason will prevail. If it does not then there is a very real danger that the whole human enterprise will fail.

In a world in which there exist many nonliterate peoples, and many quasi-literate peoples, who are destined increasingly to feel the influence of the technologically developed peoples of the world, it is helpful to understand that the social changes which it is possible for such peoples to undergo without injury to themselves can proceed at quite a rapid rate.

It is also important to understand that great care and study of the conditions are necessary before such programs of change are undertaken. Frederick the Great had to persuade his peasants to eat potatoes by sending uniformed gendarmes among them. Count Rumford (1753–1814), head of the ordnance and army of the Elector of Bavaria, had to force soldiers to cultivate and consume potatoes so that they acquired the habit and carried it with them to their people after they left the army. In the matter of food, people are particularly apt to be conservative, for the most irrational reasons, as case report after case report from the field has in recent times shown us. After the greatest trouble had been taken to persuade the Spanish-American farmers living in the Rio Grande Valley of New Mexico to substitute the growing of hybrid corn for their own poor variety, the whole scheme foundered, much to the distress of the United States Department of Agriculture, on the simple fact that the women didn't like the new corn. The men found it easier in every way to grow, the crop gave three times the yield of the native corn, the quality was superior, but all the wives had complained from the first. Its texture was wrong; it didn't hang well together for tortillas; the tortillas came out the wrong color; and so on. So the inferior native corn was replanted and the superior hybrid corn abandoned. Domestic harmony was restored; tortillas looked and tasted as they always had; and once again custom declared itself king.

The moral of this story is that when one is proposing to help a people improve its lot by changing its habits, it is first necessary to make a thorough study of the manner in which this might best be done. In the case cited above, the anthropologist would have suggested the following general steps:

1. Trial of several varieties of hybrid corn, with full recognition on the part of everyone that this was a test to discover which corn the people liked best.
2. Testing to see how the corn selected really fitted into the culture patterns.
3. Continued working with the farmers in order to see that they fully convinced themselves of the advantages of the new corn.
4. Continued contact to obviate all difficulties and to make any modifications that might be called for.

By such means, the problem which did arise might have been forestalled and met, and the society as a whole benefited.

Aware of the contribution which the anthropologist can make toward the solution of such problems, the United Nations and the state departments of many countries are now increasingly calling upon the services of anthropologists. Governments are not only calling upon anthropologists to make studies of nonliterate peoples who will soon be coming under the influences of civilization, as in South America and in India, but they are also being called upon to make studies of peoples who have already been affected by the technologically more advanced societies. And the study of such peoples is not one iota less important than the study of nonliterate peoples.

The phenomenon of cultural transmission in process from one cultural context to another is known as *acculturation,* and the study of acculturation throws valuable light upon the physiology, as it were, of culture and of cultural change. The manner in which one culture influences another is no longer merely a matter of theoretical interest, but of the greatest practical value and importance. For if the peoples of the world are to come closer together, the processes that accomplish this with the greatest benefit accruing to all must be understood. This means that the processes of acculturation, of association, and of interaction between differing peoples must be studied, analyzed, scanned, and predicted. Studies of this kind have been under way for some years, and there cannot be the least doubt that they will prove of the utmost value when they come to be applied—as some of them have already—to the solution of practical problems. It is, for example, a matter of official record that studies made by anthropologists of Japanese culture and character before the conclusion of World War II greatly helped the U.S. occupying forces of Japan. It was on the recommendation of anthropologists that the Japanese Emperor was not deposed, because as the symbol of peace for which he stood to his people, his word would be sufficient to cause all Japanese fighting men to lay down their arms. And this proved to be the case. The United Nations and the Point Four program have each employed anthropologists to deal with all those problems which fall within their competence.

In a world torn by racial misunderstanding and conflict, the contribution the anthropologist has to make is fundamental. As a student of man's mental processes, in all the forms in which he finds them, in living nonliterate and literate cultures, the anthropologist is in a position to put the problems of "race" in their proper perspective and to return the soundest possible answers—where answers are available. Because of his varied field experience with innumerable different peoples, the anthropologist is able to apply the results of that experience to the amelioration and solution of "race problems."

Anthropologists are by no means agreed that there do not exist some differences in the mental capacities of some human groups as compared with others. But they are agreed that if such differences do exist, no matter what their nature, equality as an ethical principle in no way depends upon the assertion that human beings or human groups are, in fact, equal in endowment. The individual, no matter what ethnic group he may belong to, has to be considered as a person in his own right, if for no other reason than that he is a member of the human species. The anthropologist in the field knows better than anyone else possibly can that in every nonliterate group are to be found individuals of outstanding as well as mediocre ability—just as they are to be found in literate societies. And he also knows, and hopes to convince others, that the only reasonable way in which to discover what an individual's abilities are, is not to begin with a private or public prejudice, but to treat the individual as a person and provide him with all the opportunities which will enable him to realize his potentialities. This is not only the most reasonable approach to the whole subject of race, but the scientific one, and the one that happens to correspond with the only possible democratic approach.

In discussing so-called racial differences, this book has emphasized that in the vast majority of traits, physical and mental, the likenesses far outweigh the differences. In the course of the evolution of man, the principal complex of traits which have been at the highest premium are those which are still valued by all societies and by most individuals: the capacity to get along with others, the ability to cooperate, the wisdom to know when to hold one's tongue and when to loose it; in other words, the general trait which may be called plasticity or educability. This trait is the one which natural selection and social selection has most favored in all human societies throughout the long, secular period of man's evolution. It is unlikely that in the course of that evolution any particular mental traits were ever selected anywhere nearly as consistently as the trait of plasticity has been. It also appears unlikely that, with the great history of racial mixture which mankind as a whole has behind it, and with the exchange and intermixture of genes that has resulted, there can have been any driftage of genes causing significant differences to appear in mental ability. If driftage can be eliminated—and it appears that it can be because of the racial mixing that has occurred—from the history of the development of man's genetic potentialities for mental ability, it would seem highly unlikely that there really exist any significant differences in mental capacity for the different ethnic groups of mankind. This is not to deny that such differences *may* exist, but it is to assert that if such differences do exist, those best able to demonstrate their existence, namely, the anthropologists, have been unable to find any evidence for them.

As the UNESCO *Statement on Race* puts it:

It often happens that a national group may appear to be charac-terized by particular psychological attributes. The superficial view would be that this is due to race. Scientifically, however, we realize that any common psychological attribute is more likely to be due to a common historical and social background, and that such at-tributes may obscure the fact that, within different populations consisting of many human types, one will find approximately the same range of temperament and intelligence.

Indeed, it is the usual finding that when adequate opportuni-ties for the development of capacities are made available to members of different ethnic groups and their environments are equalized as much as possible, intelligence scores become equalized also. The achievements of nonliterate peoples, as the anthropologist in the field has come to know them, have usually evoked in him feelings of the highest respect.

If there are any lessons to be learned from what anthropology has to teach us, perhaps the most important of them is that beneath all the differences in customs and body-form which human beings display there exists a fundamental likeness and kinship, a likeness and kinship which far outweigh all the differences.

If some groups of humanity are culturally more advanced in some respects than others, it is because their opportunities have been greater, and not because of some supposed innate superi-ority. No group of human beings is of less value in the balance of humanity than any other, for all groups of human beings possess the potentialities for development which, under the proper environmental stimulation, would enable them to con-tribute maximally to the achievement of humanity. Humanity, let it always be remembered, is not a gift but an achievement. And to achieve humanity opportunities are necessary.

The major manifestations of culture, art, religion, mythology, philosophy, literature, music, science, and technology, have varied enormously in different parts of the world and at differ-ent periods. Yet all cultures have had certain aims in common: to enable people to live and work together with the minimum of friction, to give people a feeling of security, a sense of belonging to a community, and a sense of purpose in living. As cultures advanced in civilization there would be added to these aims, the desire to make life cleaner, brighter, kindlier, easier, and more graceful. If we judge various cultures by their manifestations we are often bewildered by their differences. If we judge them by their aims we perceive their essential unity—a unity forged by needs and aspirations that are common to all mankind.

Every culture can be regarded as the historic result of a people's attempt to adjust itself to its environment. Before the advent of the twentieth century, that environment was usually narrowly bounded. Today the boundaries that formerly separated people are crumbling before our eyes. Mankind is moving —in spite of occasional appearances to the contrary—toward unity without uniformity, toward the condition in which the differences that today separate men will grow to be regarded, not as causes for suspicion, fear, and discrimination, but as no more important than the differences which exist between the different members of the same family. To that end, among others, the service of the anthropologist is dedicated—to make the world safe not only for democracy, but for the preservation of diversity.

FOR FURTHER READING

AGGRESSION AND WAR

Carthy, J. D., and F. J. Ebling (eds.). *The Natural History of Aggression*. Academic Press, London and New York, 1964.
A stimulating symposium on animal and human aggression.

Dollard, John, *et al. Frustration and Aggression*. Yale University Press, New Haven, 1939.
A classic work on the causes of aggressive behavior.

Fried, Morton, Marvin Harris, and Robert Murphy (eds.). *War: The Anthropology of Armed Conflict and Aggression*. Natural History Press, Garden City, New York, 1968.
A most valuable symposium on the nature, causes, and prevention of war.

Montagu, Ashley (ed.). *Man and Aggression*. Oxford University Press, New York, 1968.
A critical examination, by fourteen authorities, of the claims of some recent writers that man is innately aggressive, claims that are found wanting.

Scott, John Paul. *Aggression*. University of Chicago Press, Chicago, 1958.
An excellent account of our knowledge about aggressive behavior in modern society, its causes, control, and consequences.

THE APES

Altmann, Stuart A. (ed.). *Social Communication Among Primates*. University of Chicago Press, Chicago, 1967.
Among aspects of primate behavior discussed are dominance, aggression, subordination, social signaling, reproductive behavior, social dynamics, and the neurological and physiological correlates of social behavior.

Buettner-Janusch, John (ed.). *Evolutionary and Genetic Biology of Primates.* 2 vols. Academic Press, New York, 1963.
A review of current research on many different aspects of primate biology.

DeVore, Irven (ed.). *Primate Behavior.* Holt, Rinehart & Winston, New York, 1965.
Field studies of monkeys and apes, fascinatingly reported.

Morris, Desmond (ed.). *Primate Ethology.* Aldine, Chicago, 1967.
Invaluable studies on the behavior of monkeys and apes.

Morris, Ramona, and Desmond Morris. *Men and Apes.* McGraw-Hill, New York, 1966.
An excellent account of the historical development of our knowledge of the apes.

Napier, John R., and Prue Napier. *A Handbook of Living Primates.* Academic Press, New York, 1967.
A comprehensive catalogue of facts on primate biology and behavior.

Reynolds, Vernon. *The Apes.* Dutton, New York, 1967.
An excellent account of the history and world of the great apes and the gibbon.

Schaller, George B. *The Mountain Gorilla.* University of Chicago Press, Chicago, 1963.
————. *The Year of the Gorilla.* University of Chicago Press, Chicago, 1966.
The two best books on the gorilla in its native habitat.

Schrier, Allan M., Harry F. Harlow, and Fred Stollnitz (eds.). *Behavior of Nonhuman Primates.* 2 vols. Academic Press, New York, 1965.
Modern research trends in the study of primate behavior.

Southwick, Charles H. (ed.). *Primate Social Behavior.* Van Nostrand, Princeton, N.J., 1963.
A helpful volume of readings, mainly on the behavior of monkeys.

Yerkes, Robert M., and Ada W. Yerkes. *The Great Apes.* Yale University Press, New Haven, Conn., 1929.
The classic work, of enduring interest and value, on the behavior of the chimpanzee, orangutan, and gorilla.

APPLIED ANTHROPOLOGY

Foster, George M. *Traditional Cultures: And the Impact of Technological Change.* Harper & Row, New York, 1962.

Applied anthropology in action as exemplified by the approaches made in the field to the fitting of newly developing countries to technological and ideological change.

Gillin, John (ed.). *For a Science of Social Man*. Macmillan, New York, 1955.
The contributions that psychology, sociology, and anthropology can make towards a social science of man.

Mead, Margaret (ed.). *Cultural Patterns and Technical Change*. Mentor Books, New American Library, New York, 1954.
Originally published by UNESCO, this volume provides a fascinating series of accounts of different cultures in process of technical change.

Spicer, Edward H. (ed.). *Human Problems in Technological Change*. Russell Sage Foundation, New York, 1952.
An invaluable compendium on the problems of technological change in nonliterate cultures.

Thompson, Laura. *Toward a Science of Mankind*. McGraw-Hill, New York, 1961.
A stimulating discussion of anthropology as a unifying science of mankind.

ARCHEOLOGY

Brothwell, Don, and Eric Higgs (eds.). *Science in Archaeology*. Basic Books, New York, 1963.
On the methods and techniques of archeology, by forty-five authorities.

Caldwell, Joseph R. (ed.). *New Roads to Yesterday*. Basic Books, New York, 1966.
An illuminating series of studies on Old and New World archeology.

Hole, Frank, and Robert F. Heizer. *An Introduction to Prehistoric Archeology*. 2nd ed. Holt, Rinehart & Winston, New York, 1969.
The best general introduction to the history, method and theory of archeology.

Oakley, Kenneth. *Frameworks for Dating Fossil Man*. Aldine, Chicago, 1964.
A fundamental work which correlates the relevant data for both archeological and geological dating of fossil remains, for Middle and Upper Paleolithic sites in Europe, the Near East, and Africa.

Semenov, S. A. *Prehistoric Technology*. Barnes & Noble, New York, 1964.

A fascinating experimental study of the techniques employed by prehistoric man in the manufacture of his tools.

Willey, Gordon R. *An Introduction to American Archaeology*. Vol. 1. *North and Middle America*. Prentice-Hall, Englewood Cliffs, N.J., 1966.
The first and most thoroughly informative cultural history of Pre-Columbian America.

Wormington, H. M. *Ancient Man in North America*. Fourth ed. Denver Museum of Natural History, Denver, 1957.
An excellent and exhaustive account of the archeology of North America.

ART

Adam, Leonhard. *Primitive Art*. Third ed. Penguin Books, Baltimore, 1954.
A survey of the art of nonliterate peoples from the Paleolithic to modern times.

Bandi, Hans-George, and Johannes Maringer. *Art in the Ice Age*. Frederick A. Praeger, Inc., New York, 1953.
A beautifully illustrated account of prehistoric art.

Fraser, Douglas. *Primitive Art*. Doubleday, New York, 1962.
A beautifully illustrated account of the art of nonliterate peoples.

———— (ed.). *The Many Faces of Primitive Art*. Prentice-Hall, Englewood Cliffs, N.J., 1966.
A critical anthology, by eleven authorities, on the interpretation of nonliterate art.

Giedion, S. *The Eternal Present: The Beginnings of Art*. Pantheon Books, New York, 1962.
A magnificently illustrated study of prehistoric art.

Graziosi, Paolo. *Paleolithic Art*. McGraw-Hill, New York, 1960.
The most complete study of the art of the Old Stone Age available.

Laming, Annette. *Lascaux*. Penguin Books, Baltimore, 1959.
A valuable study of the Upper Paleolithic paintings and engravings in the famous Lascaux cave.

Leroi-Gourhan, André. *Treasures of Prehistoric Art*. Harry N. Abrams, Inc., New York, 1967.
Much more than the title implies. A magnificent original study of the meaning of prehistoric art.

Miller, M. D., and F. Rutter. *Child Artists of the Australian Bush*. Harrap & Co., London, 1952.

The story of the discovery of the astonishing artistic ability of Australian aboriginal children, with many supporting illustrations of their work.

Morris, Desmond. *The Biology of Art*. Alfred A. Knopf, Inc., New York, 1962.
A study of the picture-making behavior of the great apes and its relationship to human art.

Roy, Claude. *The Art of the Savages*. Arts Inc., New York, 1961.
A superb little book on the art of nonliterate peoples.

Ucko, Peter J., and Andrée Rosenfeld. *Palaeolithic Art*. McGraw-Hill, New York, 1967.
A superlatively good book, presenting a thoroughgoing scientific interpretation of paleolithic art. Excellently illustrated.

Wingert, Paul S. *Primitive Art*. Oxford University Press, New York, 1962.
An excellent study of the styles and traditions of the art of nonliterate peoples.

CULTURE

Benedict, Ruth. *Patterns of Culture*. Mentor Books, New American Library, New York, 1946.
A classic discussion of the meaning of culture as illustrated by four nonliterate cultures.

Bohannan, Paul, and Fred Plog (eds.). *Beyond the Frontier*. Natural History Press, Garden City, New York, 1967.
On the effect of different cultures upon one another.

Childe, V. Gordon. *Man Makes Himself*. Mentor Books, New American Library, New York, 1952.
———. *What Happened in History*. Penguin Books, Baltimore, 1945.
Two books by a great English archeologist which give an authoritative and well-told account of man's progress through the ages.

Honigmann, John J. *Culture and Personality*. Harper & Row, New York, 1954.
———. *Personality in Culture*. Harper & Row, New York, 1967.
Two excellent surveys of the available materials dealing with the interactions between personality and culture.

Hsu, Francis L. K. (ed.). *Aspects of Culture and Personality*. Abelard-Schuman, New York, 1954.
A valuable symposium on the relation between the socialization process and the development of personality.

Kluckhohn, Clyde. *Culture and Behavior*. Free Press, New York, 1962.
A working anthropologist's essays on various aspects of culture.

Kroeber, A. L. *The Nature of Culture*. University of Chicago Press, Chicago, 1952.
An admirable series of studies on the various facets of culture.

————, and Clyde Kluckhohn. *Culture: A Critical Review of Concepts and Definitions*. Papers of the Peabody Museum of Archaeology and Ethnology, Harvard University, Cambridge, Mass., vol. 47, no. 1, 1952.
An invaluable discussion and examination of the concepts of culture held at different times by different writers.

Mead, Margaret. *New Lives for Old: Cultural Transformation—Manus, 1928–1953*. William Morrow & Co., New York, 1956; Mentor Books, New American Library, New York, 1961.
An exciting restudy of the Manus Islanders after they had stepped from the Stone Age into the Air Age, showing, among other things, how speedily such a change can be made.

Schusky, Ernest L., and T. Patrick Culbert. *Introducing Culture*. Prentice-Hall, Englewood Cliffs, N.J., 1967.
A good elementary introduction to the meaning of culture.

White, Leslie A. *The Science of Culture*. Farrar, Straus & Co., New York, 1949.
————. *The Evolution of Culture*. McGraw-Hill, New York, 1959.
Two fundamental and highly original studies.

ECONOMICS

Dalton, George. *Tribal and Peasant Economies*. Natural History Press, Garden City, New York, 1967.
An anthology of studies on economic anthropology.

Herskovits, Melville J. *Economic Anthropology*. Alfred A. Knopf, Inc., New York, 1952.
On the economics of nonliterate peoples.

EDUCATION

Gruber, Frederick C. (ed.). *Anthropology and Education*. University of Pennsylvania Press, Philadelphia, 1961.
Anthropologists discuss education in contemporary pluralistic societies.

Hambly, W. D. *Origins of Education Among Primitive Peoples*. Macmillan, London, 1926.
The basic book on the subject.

Spindler, George D. (ed.). *Education and Culture.* Holt, Rinehart & Winston, New York, 1963.

A broad and fascinating survey of anthropological approaches to education.

GOVERNMENT AND POLITICS

Cohen, Ronald, and John Middleton (eds.). *Comparative Political Systems.* Natural History Press, Garden City, New York, 1967.

Twenty studies in the politics of pre-industrial societies.

Mair, Lucy. *Primitive Government.* Penguin Books, Baltimore, 1966.

A fine discussion of the law of nonliterate peoples.

Schapera, Isaac. *Government and Politics in Tribal Societies.* Watts & Co., London, 1956.

An illuminating study based on the native peoples of South Africa.

HISTORY OF ANTHROPOLOGY

Daniel, Glyn E. *A Hundred Years of Archaeology.* Macmillan, New York, 1950.

The development of archeology from 1840 to 1939.

Haddon, A. C. *History of Anthropology.* Watts & Co., London, 1949.

An excellent brief introduction to the history of anthropology.

Harris, Marvin. *The Rise of Anthropological Theory.* Thomas Y. Crowell, New York, 1968.

A splendid history of theories of culture. Highly readable.

Hays, H. R. *From Ape to Angel.* Alfred A. Knopf, Inc., New York, 1958. Paperback: G. P. Putnam's Sons, New York, 1964.

A very readable informal history of social anthropology.

Hodgen, Margaret T. *Early Anthropology in the Sixteenth and Seventeenth Centuries.* University of Pennsylvania Press, Philadelphia, 1964.

Very readable.

Kardiner, Abram, and Edward Preble. *They Studied Man.* World Publishing Co., Cleveland and New York, 1961. Paperback: New American Library, New York, 1963.

The story of ten great innovators in the history of anthropology.

Lowie, Robert H. *The History of Ethnological Theory.* Farrar & Rinehart, New York, 1937.

The standard work on the subject up to the year 1937.

Penniman, T. K. *A Hundred Years of Anthropology*. Macmillan, New York, 1952.
An excellent review of the history of anthropological thought.

Slotkin, J. S. (ed.). *Readings in Early Anthropology*. Aldine, Chicago, 1965.
Readings in the early writings on anthropology from the earliest times to the end of the eighteenth century.

INNOVATION
Barnett, H. G. *Innovation: The Basis of Cultural Change*. McGraw-Hill, New York, 1955.
An original and stimulating work on the development of novelty in different societies.

Rogers, Everett M. *Diffusion of Innovations*. Free Press, New York, 1962.
Primarily a sociopsychological approach to diffusion, throwing much light on the processes which lead to the adoption of new ideas.

ON THE INTERRELATIONSHIPS BETWEEN CULTURAL AND PHYSICAL EVOLUTION
Alland, Jr., Alexander. *Evolution and Human Behavior*. Natural History Press, Garden City, New York, 1967.
An excellent discussion of man's physical and cultural evolution.

LaBarre, Weston. *The Human Animal*. University of Chicago Press, Chicago, 1954.
A most interesting discussion of man from the unified standpoint of the cultural and physical anthropologist and depth psychologist.

Montagu, Ashley (ed.). *Culture and the Evolution of Man*. Oxford University Press, New York, 1962.
———. *Culture: Man's Adaptive Dimension*. Oxford University Press, New York, 1968.
Two books devoted to the reciprocal interaction of genetic constitution and culture in man's evolution.
———. *The Human Revolution*. Bantam Books, New York, 1967.
Cultural and genetic interaction in the evolution of man.

Spuhler, J. N. (ed.). *The Evolution of Man's Capacity for Culture*. Wayne State University Press, Detroit, 1959.
Six splendid essays on the interactive effects of man's genetic and cultural history.

INTRODUCTORY AND METHODS

Bartlett, F. C. *et al.* (eds.). *The Study of Society: Methods and Problems.* Kegan Paul, London, 1939.
What the behavioral sciences, and particularly anthropology, have done and can do.

Cohen, Yehudi A. (ed.). *Man in Adaptation.* 2 vols. Aldine, Chicago, 1968.
An excellent introductory reader. The first volume deals with the biosocial background of man's adaptation, and the second volume with the cultural present.

Epstein, A. L. (ed.). *The Craft of Social Anthropology.* Barnes & Noble, New York, 1967.
The principles of systematic field work in anthropology.

Hoebel, E. Adamson. *Anthropology: The Study of Man.* Third ed. McGraw-Hill, New York, 1966.
An excellent general introduction to anthropology.

Kluckhohn, Clyde. *Mirror for Man: The Relationship of Anthropology to Modern Life.* McGraw-Hill Paperbacks, New York, 1963.
A fine introduction to what the anthropologist tries to do.

Notes and Queries on Anthropology. Sixth ed. Royal Anthropological Institute, London, 1959.
A useful practical handbook which is informative and helpful to the general reader, and to anyone contemplating fieldwork an indispensable *vade mecum.*

Tax, Sol. (ed.). *Horizons of Anthropology.* Aldine, Chicago, 1964.
A highly readable general overview of anthropology, by anthropologists still in their thirties.

————, *et al.* (eds.). *An Appraisal of Anthropology Today.* University of Chicago Press, Chicago, 1953.
Covering all fields of anthropology.

KINSHIP AND MARRIAGE

Fox, Robin. *Kinship and Marriage.* Penguin Books, Baltimore, 1967.
The best book on the subject.

LANGUAGE

Hymes, Dell (ed.). *Language in Culture and in Society.* Harper & Row, New York, 1964.
A reader in linguistics and anthropology.

Landar, Herbert. *Language and Culture.* Oxford University Press, New York, 1966.

An introduction to the major issues in contemporary linguistics.

LAW

Bohannan, Paul (ed.). *Law and Warfare.* Natural History Press, Garden City, New York, 1967.

Anthropological studies on law, raiding, conflict, and warfare.

Hoebel, E. Adamson. *The Law of Primitive Man.* Harvard University Press, Cambridge, Mass., 1954.

Describes and analyzes the legal culture of seven selected non-literate societies, and presents a theory of law and society.

LITERATURE

Astrov, Margot (ed.). *The Winged Serpent.* The John Day Co., New York, 1946.

An anthology of American Indian prose and poetry.

Cushing, Frank H. *Zuñi Folk Tales.* Alfred A. Knopf, Inc., New York, 1931.

Magnificent examples of the literature of a nonliterate people.

Greenway, John (ed.). *The Primitive Reader.* Folklore Associates, Hatboro, Pa., 1965.

An anthology of myths, tales, songs, riddles, and proverbs of aboriginal peoples around the world.

————. *Literature Among the Primitives.* Folklore Associates, Hatboro, Pa., 1964.

A stimulating study of the anthropology of literature.

Rothenberg, Jerome (ed.). *Technicians of the Sacred.* Doubleday, New York, 1968.

A range of poetries from Africa, America, Asia, and Oceania.

Trask, Willard R. (ed.). *The Unwritten Song.* 2 vols. Macmillan, New York, 1966/67.

Poetry of the nonliterate and traditional peoples of the world.

MAGIC AND WITCHCRAFT

Middleton, John (ed.). *Magic, Witchcraft, and Curing.* Natural History Press, Garden City, New York, 1967.

Studies on magic, witchcraft, sorcery, divination, and curing.

MAN'S EVOLUTION

Blum, Harold F. "Does the Melanin Pigment of Human Skin Have Adaptive Value?" *Quarterly Review of Biology,* vol. 36, 1961, pp. 50–63.

A stimulating paper questioning current views concerning the adaptive value of skin pigment, and the differences in skin color.

Brace, C. Loring, and Ashley Montagu. *Man's Evolution*. Macmillan, New York, 1968.
An introduction to the study of man's evolution.

Campbell, Bernard. *Human Evolution*. Aldine, Chicago, 1966.
An excellent introduction to the study of man's adaptations.

Day, Michael. *Guide to Fossil Man*. World Publishing Co., Cleveland and New York, 1966.
A most helpful guide and handbook of human paleontology.

Dobzhansky, Theodosius. *Evolution, Genetics and Man*. John Wiley & Sons, Inc., New York, 1955.
An authoritative work on the mechanisms of evolution with especial reference to man.

———. *Mankind Evolving*. Yale University Press, New Haven, Conn., 1962.
On the evolution of man and the genetic basis of culture.

Harrison, G. A., J. S. Weiner, J. M. Tanner, and N. A. Barnicot. *Human Biology*. Oxford University Press, New York, 1964.
An introduction to human biology, evolution, variation, and growth.

Howell, F. Clark. *Early Man*. Time Inc., New York, 1965.
A splendidly illustrated account of the evolution of man.

Howells, William (ed.). *Ideas on Human Evolution*. Harvard University Press, Cambridge, Mass., 1962.
Twenty-eight essays representing recent significant contributions to the unraveling of man's evolution.

———. *Mankind in the Making*. Doubleday, New York, 1967.
A highly readable account of the origin and evolution of man.

Laughlin, W. S., and R. H. Osborne (eds.). *Human Variation and Origins*. W. H. Freeman & Co., San Francisco, 1967.
A collection of twenty-seven articles from *Scientific American* on human biology and evolution.

Montagu, Ashley. *An Introduction to Physical Anthropology*. Third ed. Charles C Thomas, Springfield, Ill., 1960.
For the layman and beginning student, an introduction to the origin, evolution, and development of man as a physical organism.

Washburn, Sherwood L. (ed.). *Classification and Human Evolution*. Aldine, Chicago, 1963.
An invaluable work by leading authorities on the evolution of man and the classification of fossil primates.

Washburn, Sherwood L. (ed.). *Social Life of Early Man.* Quadrangle Books, Chicago, 1961.
A valuable symposium on a much-neglected subject.

MUSIC

Bowra, C. M. *Primitive Song.* World Publishing Co., Cleveland and New York, 1962.
An illuminating study of the songs of nonliterate peoples.

Nettl, Bruno. *Theory and Method in Ethnomusicology.* Free Press, New York, 1964.
A fundamental work on the approach to the study of the music of nonliterate and folk cultures.

MYTHOLOGY

Marriott, Alice, and Carol K. Rachlin. *American Indian Mythology.* Thomas Y. Crowell, New York, 1968.
The myths and legends of more than twenty American Indian tribes.

Middleton, John (ed.). *Myth and Cosmos.* Natural History Press, Garden City, New York, 1967.
Readings on the relationships between creation myths and folklore, and the human behavior of nonliterate societies.

PHILOSOPHY

Radin, Paul. *Primitive Man as Philosopher.* Appleton-Century, New York, 1927. Reprinted, Dover Publications, New York, 1958.
An extremely interesting and readable discussion of nonliterate man as a thinker.

RACE

Asimov, Isaac, and William C. Boyd. *Races and People.* Abelard-Schuman, New York, 1955.
A clear account of the mechanisms of inheritance as they relate to the differentiation of man.

Dunn, L. C., and Theodosius Dobzhansky. *Heredity, Race and Society.* Mentor Books, New American Library, New York, 1952.
A scientific explanation of human differences.

Mead, M., T. Dobzhansky, E. Tobach, and R. E. Light (eds.). *Science and the Concept of Race.* Columbia University Press, New York, 1968.
By far the best book on the subject by twenty contributors drawn from anthropology, biology, ethnology, genetics, physiology, and psychology.

Montagu, Ashley. *Man's Most Dangerous Myth: The Fallacy of Race*. Fourth ed. World Publishing Co., Cleveland and New York, 1964.

A full discussion of the subject.

——— (ed.). *The Concept of Race*. Free Press, New York, 1964.

A critical examination of the concept of race by ten authorities.

UNESCO. *The Race Question in Modern Science*. UNESCO, New York, 1957.

An authoritative and clear discussion of the various aspects of the race question by nine experts.

RELIGION

Howells, William. *The Heathens: Primitive Man and His Religions*. Doubleday, New York, 1948.

An able and readable account of primitive religion.

Middleton, John (ed.). *Gods and Rituals*. Natural History Press, Garden City, New York, 1967.

Readings in religious beliefs and practices.

Norbeck, Edward. *Religion in Primitive Society*. Harper & Bros., New York, 1961.

An excellent discussion of the meaning of religion in nonliterate societies.

TECHNOLOGY

Singer, Charles, E. J. Holmyard and A. R. Hall (eds.). *A History of Technology*. Vol. 1. Oxford University Press, New York, 1954.

An authoritative account by experts of the technologies of man from the earliest times to the fall of ancient empires.

THEORY

Bidney, David. *Theoretical Anthropology*. Columbia University Press, New York, 1953.

A broad and penetrating survey of the ideas of anthropology.

Manners, Robert O., and David Kaplan (eds.). *Theory in Anthropology*. Aldine, Chicago, 1968.

A fine sourcebook on anthropological theory.

Nadel, S. N. *The Foundations of Social Anthropology*. Cohen & West, London, 1951.

A brilliant book on the methods and ideas of anthropology.

INDEX

NOTE: *Italicized page numbers refer to illustrations.*

of man, 100-3, 237; and environment, 5, 8, 52, 100-1, 116-17; evolution and development of, 49, 52, 119-20, 237-38; and human needs, 116-23; and individual, 120-21; and language, 124-31; and personality, 6

Cultures, acculturation, 214, 234-35; studied by anthropologists, 5-7, 231-35; and archeologists, 7-8; culture areas of Africa, 109; of Asia, 110; of North America, 88, 112; of South America, 90; Aurignacian, 221; creation stories, 207; a culture defined, 101; dance, 229-30; death, 204; food habits, 134, 234; imaginative life, 197; industry, 151; kinship behavior, 178; languages, 124; literature, 225; mythology and folklore, 197-99; Neolithic in Egypt, 140-42; Oldowan, 46; population, 97; "race," 235; religion, 194; sex-differentiation, 161; transference or combination as common cultural process, 230; variety of cultures, 102-3, 237

Cuneiform, Sumerian, 125, 132
Curtis, G. H., 44
Cushing, F. H., 225

Dahomeans, 169
Dance, 2, 120, 159, 219, 225, 226, 229-30
Dart, Raymond, 42, 48
Darwin, Charles, 11, 94-95
Dating methods, *see* Potassium-argon, Radiocarbon, *and* Uranium transformation
Dayaks, 107 (Table 6), 149
Democritus, 209
Descent, 158, 159, 176, 186; *see also* Matrilineal *and* Patrilineal
Dieri, 215-16
Differentiation of man, 93-103
Diffusion, 199, 218
Dinaric, 111
Divination, 194
Division of labor between sexes, 134, 149, 161
Divorce, 120, 168-169
Diyarbakir, 217
Domestication of plants and animals, 137-43, *144,* 148, 150
Dravidian, 111
Drawing, 120, 132, 135, 219-24, 229
Dress, *see* Clothing and dress
Dryopithecus, 18-19 (Table 2), 33

Dubois, Eugene, 54, 62
Durkheim, Emile, 192
Durukulis, 23
Dutch, 123
Dwellings, 46, 49, 147; *see also* Shelter

Education (educability of man), 4, 52, 102, 118, 131, 236
Egypt (Egyptian), cranial capacity, 107 (Table 6); culture stages, 80-83 (Table 4); hieroglyphics, 132; incest, 163; industry, 151; Neolithic, 140-42, 217; pottery dish, *142;* transport, 152
Elopement, 168
Elwin, Verrier, 225
Endogamy, 163, 165, 175
England (English), 6, 76, 79, 99, 107 (Table 6), 130, 131, 134, 188
Environment, adaptation and adjustment to, 11, 39, 52-53, 95, 102, 173, 182, 200, 232, 238; and culture, 5, 8, 52, 100-1, 116-18; and heredity, 120; and intelligence, 200, 237; interaction with, 3-4; and land-sale, 156; responses to challenges of, 50, 122, 137; and shelters, 147; and time-reckoning, 213
Erect or upright posture, 26, 32, 35, 39-41, 50, 55, 67, 101
Eskimos, 110 (map), 174; aggression, lack of, 122, 149; bolas, 58; clothing, 145; cranial capacity, 107 (Table 6); crime, view of, 187; Dance Song, 226; family life, 173; fire-making, 210; government, 185; ethnic groups, 114, 115; hunting methods, 134, 136; illegitimacy, 177; immortality doctrine, 204; language, 130; mores, 183; origin of name, 182; population size, 97; property rights, 158; shelters, 146-47; territoriality, lack of, 123; time-reckoning, 213; tools, 217; transportation devices, 152
Ethnic groups, 9, 93-95, 101-2, 104-15, 164, 236-37
Ethnomusicology, 227
Europe, Barbary ape, 25; Caucasoids, 106, 111; cave art, 224; cave dwellers, modern, 146; cowrie shells, 156; cranial capacity, 107 (Table 6); culture stages, 80-83 (Table 4); dowry, 167; *Dryopithecus,* 33; earliest

also *Sinanthropus*, 58-60, *58*, 210
Homo erectus (Pithecanthropus) robustus, 56, *75*, 76, 83 (Table 4), 107 (Table 6)
Homo habilis (habilines), 44, 48, 49, 56, 83 (Table 4), 107 (Table 6)
Homo sapiens, 1-2, 4, 16, 18-19 (Table 2), 62, 80-83 (Table 4), 94
Homo sapiens fossilis, *74*
Hong Kong, 123
Honigmann, John, 6
Hopi Indians, 88-89 (map), 158, 175
Hottentots, 47, 107 (Table 6), 108, 109, 158, 182-83, 227
Hsu, Francis, 6
Hunt (hunting), and australopithecines, 46, 57; and evolution, 39, 49-52, 100-1, 122; lower hunter stage, 97; mana in, 192; proficiency, 203; rights, 158, 183; signs, 142; and speech, 127-28
Hunting peoples, 123, 134-37, 140, 141, 150, 152, 158, 174
Huxley, T. H., 9-10
Hybridization and interbreeding, 71, 93, 94, 97-98, 106

Ifaluk, 122
Immortality, 204
Implements, *see* Tools and implements
Incas, 146, 217
Incest, 163-64, *165*, 188
India, Buddha, 2; Caucasoids, 108, 111; culture areas and tribes of, 110-11 (map); *Dryopithecus*, 33; folklore, 198; folk-songs, 225; Gandhi, 2; Indians, (Asiatic), 93, 123, 218; Indo-Dravidians and Pre-Dravidians, 111; lemur, 16, 17; Muria, 162; studies of non-literate peoples, 234; skin cancer, 95; Todas, 162; tree shrew, 16
Indians, American, 93; areas of culture and tribes, in North and South America, 88, 90 (map); canoes, 153; Caucasoid, 89; clothing, 144-46; cranial capacity, 107 (Table 6); death, 204-5; divorce, 169; ethnic groups, 115; fire-making, 210; fishing, 134; folk tales, 198; gestures, 131; guardian spirits, 194; hunting methods, 135; joking rela-

tionships, 178; languages, 130; markets, 157; marriage practices, 166, 167; Mongoloids, 106; music, 227; philosophy, 200-1; polygyny, 166; raiding, 148-49; status of women, 169; supernatural, 193; tools, 217; writing, 132-33, *133*; *see also under tribal names*
Indies, East, 106, 115
Indo-Chinese, 123
Indo-Germanic, 130
Indo-Malay, 115
Indonesia, 111, 115, 168
Indus, 151
Industry, 7, 9, 120, 149, 151
Inheritance, 94, 120, 158, 159, 168, 186
Insectivores, 16, 18-19 (Table 2)
Instincts, 50-52, 118, 122-23
Institutions, 5, 117, 118, 125, 150, 163, 170, 186
Intelligence, of aborigines, 200; in different ethnic groups, 237; in Eskimos, 130; in evolution of man, 50-52, 118, 122, 128; in hunting peoples, 136; in hybrids, 98; Neanderthal, 67; nonhuman primates, 25, 26, 27, 36; in Old Stone Age, 219; of pharaohs, 163; and science, 208, 211; *see also* Wisdom
Interglacial period, 77
Invention, 217-18
Iran, 149
Irano-Afghan Mediterranean ethnic group, 108
Iraq, 138
Iroquois, 88-89 (map), 107 (Table 6), 147, 169, 172, 185, 186, 193
Isolation, 94, 96, 97, 106
Israel, *70*, 170
Italy, 72, 131, 134

Jagga, 189
James, Henry, 129
Japan, 6, 59, 107 (Table 6), 110 (map), 111, 168, 184, 235
Jarmo, 81 (Table 4), 138, 142
Java, 44, 54, *55*, 56, 59, 60, *61*, 62, 65
Jericho, 81 (Table 4), 140
Jespersen, Otto, 126-27
Jews, 131, 166; Judaism, 195
Joking relationship, 178

Kamasian industry, 60
Kambots, 166
Kamchadales, 114, 115
Kanam man, 78-79, 84

733 -469

Tools and implements (*Cont.*)
of Solo man, 61; of Swanscombe
man, 76; as weapons, 149
Totem, 133, 176, 191-92
Totemic stones, *190*
Trade, 120, 140, 151, 153-57
Transportation, 120, 147, 151-53
Travel, 2, *85,* 120, 152-53, 198
Tree shrew, 16-17, *17*
Trobriand Islands, 165, 172, 178
Tungus, 114
Tupaias, 16-17

Uganda, 157
UNESCO *Statement on Race,* 237
United Nations, 234, 235
United States, 93, 188, 232-33;
see also America, North
Unkulunkulu, 201-2
Uranium transformation method
of dating, 18-19 (Table 2)
Urban Revolution, 151

Veddah, 107 (Table 6), 111
Victoria, Queen, 99, *100;* Victorian period, 161

Wadjak man, 62, 83 (Table 4),
107 (Table 6)
Wales, 72
Weapons, 57-58, 119, 135, 148-49,
153
Weidenreich, Franz, 54
Whites, 89, 93, 95, 96, 99, 106,
143, 200, 229

Winnebago, 88-89 (map), 167,
203, 205
Wisdom, 1-3, 157, 185, 203, 205,
236
Wishram, 157
Witchcraft, 188
Witoto, 90-91 (map), 169, 175
Women, a capital crime, 189; con-
servatism of, 234; divorce privileges, 169; feminine traits ex-
amined, 162; first gardeners and
agriculturists, 137, 150; natural
selection of, 51; in politics, 185-
86; sexual selection, 98-99; status, 150, 169; suffrage movement, 2; work and occupations
of, 149-50, 161; among Zuñi,
157
Wright, Sewall, 97
Writing, 6, 119, 125, *132, 133,*
132-33, 140-42
Würm glaciation, 82 (Table 4),
146

Yucatan, 133
Yugoslavs, 166
Yukaghirs, 167
Yurok, 88 (map), 169, 188

*Zinjanthropus (Paranthropus ro-
bustus) boisei,* 42-45, *45,* 47-
49, *84,* 107 (Table 6)
Zuñi, 88-89 (map), 156-57, 169,
194, 214
Zuñi Folk Tales (Cushing), 225